Edited by Editorial Board of UTokyo Engineering Course

Complex Function Theory

UTokyo Engineering Course/Basic Mathematics

Linear Algebra I: Basic Concepts
by Kazuo Murota and Masaaki Sugihara
ISBN: 978-981-125-702-5
ISBN: 978-981-125-797-1 (pbk)

Linear Algebra II: Advanced Topics for Applications
by Kazuo Murota and Masaaki Sugihara
ISBN: 978-981-125-705-6
ISBN: 978-981-125-798-8 (pbk)

Partial Differential Equations
by Osamu Sano
ISBN: 978-981-127-088-8
ISBN: 978-981-127-131-1 (pbk)

UTokyo Engineering Course / Basic Mathematics

Edited by Editorial Board of UTokyo Engineering Course

Complex Function Theory

Takeo Fujiwara

Professor Emeritus at The University of Tokyo, Japan

Published by

World Scientific Publishing Co. Pte. Ltd.

5 Toh Tuck Link, Singapore 596224

USA office: 27 Warren Street, Suite 401-402, Hackensack, NJ 07601

UK office: 57 Shelton Street, Covent Garden, London WC2H 9HE

and

Maruzen Publishing Co., Ltd.
Kanda Jimbo-cho Bldg. 6F, Kanda Jimbo-cho 2-17
Chiyoda-ku, Tokyo 101-0051, Japan

British Library Cataloguing-in-Publication Data
A catalogue record for this book is available from the British Library.

UTokyo Engineering Course/Basic Mathematics
COMPLEX FUNCTION THEORY

Copyright © 2023 by Takeo Fujiwara

ISBN 978-981-127-091-8 (hardcover)
ISBN 978-981-127-132-8 (paperback)
ISBN 978-981-127-092-5 (ebook for institutions)
ISBN 978-981-127-093-2 (ebook for individuals)

For any available supplementary material, please visit
https://www.worldscientific.com/worldscibooks/10.1142/13267#t=suppl

Desk Editor: Tan Rok Ting

Printed in Singapore

UTokyo Engineering Course

About This Compilation

What is the purpose of engineering education at the University of Tokyo's Undergraduate and Graduate School of Engineering? This School was established 125 years ago, therefore we feel it is an appropriate time to ask this question again. More than a century has passed since Japan embarked on a path to introduce and negotiate Western knowledge and practices. Japan and the world are very different places now, and today our university stands as a world leading institute in engineering research and education. As such, it is our duty and mission to build a firm foundation of education that will support the creation and dissemination of engineering knowledge, practices and resources. Our School of Engineering must not only teach outstanding students from Japan but also those from throughout the world. Put another way, the engineering that we teach students is not only a responsibility of this School, but an imperative placed on us by society and the age in which we live. It is in this changed context, where we have gone from follower to leader, that we present this curriculum, The University of Tokyo (UTokyo) Engineering Course. The course is a reflection of the School's desire to engage with those outside the walls of the Ivory Tower, and to spread the best of engineering knowledge to the world outside our institution. At the same time, the course is also designed for the undergraduate and graduate students of the School. As such, the course contains the knowledge that should be learnt by our students, taught by our instructors and critically explored by all.

February 2012   Takehiko Kitamori
            Dean, Undergraduate and Graduate Schools of Engineering
                                          The University of Tokyo
                                            (April 2010–March 2012)

v

UTokyo Engineering Course

The Purpose of This Publication

Modern engineering is composed of the academic discipline of fundamental engineering and the academic discipline of integrated engineering that deals with specific systems and subjects. Interdisciplinary disciplines and multidisciplinary disciplines are amalgamations of multiple academic disciplines that result in new academic disciplines when the academic pursuit in question does not fit within one traditional fundamental discipline. Such interdisciplinary disciplines and multidisciplinary disciplines, once established, often develop into integrated engineering. Moreover, the movement toward interdisciplinarity and multidisciplinarity is well underway within both fundamental engineering and advanced research.

These circumstances are producing a variety of challenges in engineering. That is, the scope of research of integrated engineering is gradually growing larger, with economics, medicine and society converging into an enormously complex social system, which is resulting in the trend of connotative academic disciplines growing larger and becoming self-contained research fields, which, in turn, is resulting in a trend of neglect toward fundamental engineering. The challenge of fundamental engineering is how to connect engineering education that is built upon traditional disciplines with that of advanced engineering research in which interdisciplinarity and multidisciplinarity is continuing at a rapid pace. Truly this is an educational challenge shared by all the top engineering schools in the world. Without having a solid understanding of engineering, however, education related to learning state-of-the-art research methodologies will not hold up. This is the dichotomy of higher education in engineering; that is, higher education in engineering simply will not work out if either side of the equation is missing.

In the meantime, the internationalization of universities is going forward in routine fashion. In fact, here at the University of Tokyo (UTokyo), one quarter of the graduate students enrolled in engineering fields are of foreign nationality and the percentage of foreign undergraduate students is expected to increase more and more. On top of that, Japan is experiencing a reduction in the population of its youth. Therefore, the time is ripe to ramp up efforts to look outside of Japan in order to secure the human resources to sustain the future of advanced science and technology here in Japan. It is clear that the internationalization of engineering education is rapidly underway. As such, the need for a curriculum that is firmly rooted in engineering knowledge needs to be oriented toward both local and foreign students.

Due to these circumstances surrounding modern engineering, we at UTokyo's School of Engineering have systematically organized an engineering curriculum of fundamental engineering knowledge that will not be unduly influenced by the times, with the goal of firmly establishing a benchmark suitable for that of the top schools of engineering of science and technology for students to learn and teachers to teach. This engineering curriculum clarifies the disciplines and instruction policy of UTokyo's School of Engineering and is composed of three layers: Fundamental (sophomores (second semester) and juniors), Intermediate (seniors and graduates) and Advanced (graduates). Therefore, this engineering course is a policy for the thorough education of the engineering knowledge necessary for forming the foundation of our doctorate program as well. The following is an outline of the expected effect of this engineering course:

- Surveying the total outline of this engineering course will assist students in understanding which studies they should undertake for each field they are pursuing, and provide an overall image by which the students will know what fundamentals they should be studying in relation to their field.
- This course will build the foundation of education at UTokyo's School of Engineering and clarify the standard for what instructors should be teaching and what students need to know.
- As students progress in their major it may be necessary for them to go back and study a new fundamental course. Therefore the textbooks are designed with such considerations as well.

- By incorporating explanations from the viewpoint of engineering departments, the courses will make it possible for students to learn the fundamentals with a constant awareness of their application to engineering.

Yasuhiro Kato, Board Chair
Yukitoshi Motome, Shinobu Yoshimura, Executive Secretary
Editorial Board of UTokyo Engineering Course

Preface

This book is an English version of "Complex Function Theory" in the University of Tokyo Engineering Course prepared in Japanese for undergraduate students. The goal of the book is to provide ease of use and applicability, not mathematical rigor. The explanations avoid abstract or too many general points as much as possible and are based on a few key examples to help university students understand and use the important properties of complex functions.

The complex function theory in this book is mainly classical mathematics, which was established in the 19th or early 20th century. We believe studying complex function theory is significant in two ways. One is that it prepares the way for applied mathematics that is used widely in specialist fields. A deeper study of differential equations and Fourier–Laplace analysis is not possible without the help of complex function theory. These are also points of entry to mathematics as a more broadly applicable tool. The other is that complex function theory is one of the classical forms of mathematics, so some aspects of it are good educational material for reconsidering the form of mathematics as an academic discipline, and from the perspective of slightly deeper immersion in specialist areas. Therefore, the complex function theory generally comes up in the second year, at the earliest, or in the third year after studying calculus and linear algebra right after entering university.

The book begins by defining complex numbers and their addition, subtraction, multiplication, and division. Next, in Chapter 2, we move on to the differentiation of holomorphic complex functions. Chapter 3 describes elementary functions and Chapter 4 advances to conformal mapping. The classification of singularities explained in Chapter 5 is very important, and then in Chapter 6 we define complex integrals and study the residues theorem. Chapter 7 describes some applications of complex integrals and discusses the regularity of functions and the equivalence of Cauchy's integral theorem. This chapter describes the Taylor and Laurent expansions in an orthodox way.

However, at the very early stage from the practical viewpoint, we often have expressed functions in the form of series, or use Riemann surface in explanations of the definition of elementary functions, because they make understanding much easier.

Chapter 8 is a revisit to the modern basis of complex integral and related topics. The analytic continuation is discussed in Chapter 9.

Chapter 11 and later chapters will cover several subjects that may not be discussed in the standard complex function theory at the undergraduate level. These are the foundations of applied mathematics that are necessary for later undergraduate and post-graduate courses. Therefore, elements of these subjects are described only piecemeal, as necessary within the standard course, and there is almost no cohesive discussion. In that sense, these chapters may have a role in providing supplementary learning material for content that should be learned within undergraduate courses. Alternatively, it may be a preparation of matters necessary in more advanced lecture courses and research and application.

I have tried to keep this textbook as self-contained as possible so that the reader does not need to refer to other books. The literature at the end of the book was used as a reference in writing this textbook. As I mentioned at the beginning, the foundation of complex function theory was a field that was almost completed in the 19th or early 20th century, and the references listed at the end of the book are relatively old. On the other hand, the old-fashioned style of these documents is specific and fairly easy to understand for beginners.

Summer 2022

Takeo Fujiwara

Contents

Chapter 1

Complex numbers and their functions

To learn various properties of functions of complex variables (we call them complex functions), we should first understand the properties of complex numbers. In this first chapter, we start by defining complex numbers and their arithmetic operations (addition, subtraction, multiplication, and division). Then we will describe the relationship between complex numbers and two-dimensional vectors. In the second half of this chapter, we will discuss the limit of a sequence and the convergence of a series of complex numbers. The idea of convergence and limit is not essentially different from the cases of real numbers. Even so, an infinite series is not only meaningful in itself but is also indispensable for understanding the property of complex functions. Most of the commonly-used elementary complex functions are defined in the form of an infinite series.

1.1 Complex numbers

1.1.1 *Definition of a complex number*

Let i be the number that becomes -1 when squared.[1] This number is called the **imaginary unit**. Thus

$$i^2 = -1. \tag{1.1}$$

This imaginary unit i is used to define a new kind of number.

Definition 1.1 (Complex number). Using a pair of two real numbers x and y, the "number " z is defined as

$$z = x + iy \tag{1.2}$$

---

[1]It may be written as j to avoid confusion with the i used to represent electric current in engineering.

1

and is called a **complex number**. Here, x is called the **real part** of the complex number z, and y the **imaginary part** of the complex number z. The complex number iy that consists of only the imaginary part is called a (pure) imaginary number.[2] The real part and imaginary part of the complex number z are written as "Re z" and "Im z", respectively.

Unless otherwise specified, we write, in this book, complex numbers as z, w and real numbers as x, y, u, v. Therefore, if we write $z = x + iy$ or $w = u + iv$, x or u are the real parts and y or v are the imaginary parts. As a pair of two real numbers x, y defines a single complex number z, it may be also written as

$$z = (x, y). \tag{1.3}$$

A complex number in which the real and imaginary parts are each 0 is called the "complex number 0".

$$z = x + iy = 0 \quad \Longleftrightarrow \quad x = 0, \quad y = 0. \tag{1.4}$$

Definition 1.2 (Equality of complex numbers). If the real parts and the imaginary parts of two complex numbers z_1 and z_2 are each equal, it is said that "the complex numbers z_1 and z_2 are equal". Conversely, two complex numbers z_1 and z_2 can only be equal, if the real parts of z_1 and z_2 are equal, and so are the imaginary parts.

$$z_1 = z_2 \Longleftrightarrow \begin{cases} \text{Re } z_1 = \text{Re } z_2 \\ \text{Im } z_1 = \text{Im } z_2 \end{cases}. \tag{1.5}$$

$z' = x - iy$, in which the sign of the imaginary part of the complex number $z = x + iy$ is reversed, is called the **complex conjugate** of z, and is expressed as $\bar{z}$.

$$\bar{z} = x - iy. \tag{1.6}$$

In the field of physics, the complex conjugate is often written as z^*.

$\sqrt{x^2 + y^2}$ for the complex number $z = x + iy$ is called the **absolute value** of z, and is written $|z|$.

$$|z| = \sqrt{x^2 + y^2}. \tag{1.7}$$

The absolute value of the complex number $z = x + iy$ is 0 only when z is 0.

$$|z| = 0 \Longleftrightarrow \begin{cases} x = 0 \\ y = 0 \end{cases}. \tag{1.8}$$

The absolute value of z and the absolute value of $\bar{z}$ are equal.

$$|z| = |\bar{z}|. \tag{1.9}$$

---

[2]Imaginary numbers were discovered by an Italian mathematician G. Cardano (1501–1576).

1.1.2 Addition, subtraction, multiplication, and division of complex numbers

So far, we have defined complex numbers but have not defined the rules for how to "calculate" them. We start by defining the arithmetic operations, *i.e.* addition, subtraction, multiplication, and division, of complex numbers, so that these calculations can be performed.

Definition 1.3 (Addition and subtraction). The addition and subtraction of two complex numbers $z_1 = x_1 + iy_1$, $z_2 = x_2 + iy_2$ are defined as

$$z_1 \pm z_2 = (x_1 \pm x_2) + i(y_1 \pm y_2) \qquad \text{(double-sign correspondence)}. \quad (1.10)$$

Definition 1.4 (Multiplication). The multiplication of two complex numbers is defined as

$$z_1 z_2 = (x_1 x_2 - y_1 y_2) + i(x_1 y_2 + y_1 x_2). \qquad (1.11)$$

This can be performed, in the same way as the multiplication of real numbers, taking care to note $i^2 = -1$, and we obtain

$$z_1 z_2 = (x_1 + iy_1)(x_2 + iy_2) = (x_1 x_2 + i^2 y_1 y_2) + i(x_1 y_2 + y_1 x_2)$$
$$= (x_1 x_2 - y_1 y_2) + i(x_1 y_2 + y_1 x_2). \qquad (1.12)$$

Definition 1.5 (Multiplicable inverse, reciprocal). When $z = x + iy \neq 0$, it is possible to define a complex number z', which satisfies a relation $zz' = 1$. This is called the multiplicable inverse of z or the reciprocal of z, and is written as z^{-1}. Specifically, it is written as

$$z^{-1} = \frac{x - iy}{x^2 + y^2} = \frac{\bar{z}}{|z|^2}. \qquad (1.13)$$

This can be confirmed immediately as $zz^{-1} = (x+iy) \cdot (x-iy)/(x^2+y^2) = 1$.

Definition 1.6 (Division). When a complex number $z_2 \neq 0$, the division z_1/z_2 is defined as follows:

$$\frac{z_1}{z_2} = z_1 z_2^{-1} = z_1 \frac{\bar{z}_2}{|z_2|^2} = \frac{z_1 \bar{z}_2}{|z_2|^2}$$
$$= \frac{(x_1 x_2 + y_1 y_2) + i(-x_1 y_2 + y_1 x_2)}{x_2^2 + y_2^2}. \qquad (1.14)$$

Theorem 1.1. *The associative law and commutative law for the addition (subtraction) of complex numbers*

$$\begin{aligned}
(z_1 + z_2) + z_3 &= z_1 + (z_2 + z_3) &\quad (\textit{Associative law}), \\
z_1 + z_2 &= z_2 + z_1 &\quad (\textit{Commutative law})
\end{aligned} \qquad (1.15)$$

and the associative law and commutative law for multiplication

$$(z_1 z_2)z_3 = z_1(z_2 z_3) \qquad (\textit{Associative law}),$$
$$z_1 z_2 = z_2 z_1 \qquad (\textit{Commutative law}) \tag{1.16}$$

are all valid. Also, the distributive law

$$z_1(z_2 + z_3) = z_1 z_2 + z_1 z_3 \qquad (\textit{Distributive law}) \tag{1.17}$$

is valid.

Using z and its complex conjugate $\bar{z}$, the absolute value can be written as

$$|z| = \sqrt{z\bar{z}}. \tag{1.18}$$

For the absolute values of the sum and product of complex numbers,

$$|z_1 + z_2| \leq |z_1| + |z_2|, \qquad |z_1 z_2| = |z_1| \cdot |z_2| \tag{1.19}$$

are valid.

The various laws (1.15)–(1.19) are easy to prove, so readers should attempt to do so.

1.2　Complex plane

1.2.1　*Complex plane and complex numbers*

A **number line** is used to represent real numbers. Since a complex number is a pair of real numbers, it can be expressed using a two-dimensional vector space.

Definition 1.7 (Complex plane). If the x coordinate and y coordinate on a two-dimensional plane are made to correspond to the real part x and imaginary part y of a complex number $z = x+iy$, the point (x, y) on the two-dimensional plane corresponds to the complex number z. Consequently, the complex number 0 corresponds to the origin of the plane. This two-dimensional plane is called a **complex plane**, or **Gauss plane**.[3] The x axis of the complex plane is called the **real axis**, and the y axis is called the **imaginary axis**.

---

[3]Johann Carl Friedrich Gauss [1777–1855] made a complex number corresponding to one point on a plane.

The complex plane can make the mutual relationships between complex numbers clearer.[4] Since the complex conjugate $\bar{z}$ has the sign of the imaginary part of z changed, the points z and $\bar{z}$ are in positions symmetric to each other with respect to the real axis. When one considers two vectors, on the two–dimensional plane, from the origin to points z_1 and z_2, the point on the plane indicated by the sum of two vectors corresponds to the complex number $z_1 + z_2$ on the complex plane.[5] The absolute value $|z|$ of the complex number z is the distance from the origin to the point $z = (x, y)$ on the complex plane.

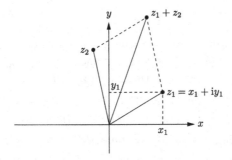

Fig. 1.1　The complex plane and the sum of complex numbers $z_1 + z_2$.

Similarly, $|z_1 - z_2|$, the absolute value of the difference between the two complex numbers z_1, z_2, is the distance between two points z_1 and z_2 on the complex plane. Considered this way, two complex numbers z_1 and z_2 satisfy the relation

$$|z_1 \pm z_2| \leq |z_1| + |z_2|. \tag{1.20}$$

This is the **triangle inequality**, which means that "the sum of two sides of a triangle is greater than or equal to the third side".

---

[4]For a real number, it is possible to describe the size relationship between two real numbers by their positions on a number line. It is not possible to relate two complex numbers using inequality signs, but they can be understood by their positional relationship on a complex plane.

[5]Complex numbers are points in real two-dimensional vector space, and their sum or difference was made to correspond to the sum or difference of two-dimensional vectors. However, the multiplication of two complex numbers does not correspond to the product (multiplication) of vectors.

1.2.1.1 *Line and circle on the two-dimensional plane*

The equation for a shape on a two-dimensional plane can be written using vectors. Therefore, the equation for a shape on the two-dimensional plane can be written using complex numbers. Let us examine mathematical expressions for shapes using complex numbers.

Example 1.1. (Equation of a straight line) A straight line on a two-dimensional plane can be written as $ax + by + c = 0$, where a, b, and c are real numbers. Therefore, a straight line on a complex plane, as $\alpha = a + ib$, $z = x + iy$, is expressed as

$$\bar{\alpha}z + \alpha\bar{z} + 2c = 0. \tag{1.21}$$

Example 1.2. (Equation of a circle) The equation for a circle on a two-dimensional plane with its center at the point (x_0, y_0) and radius a is $(x - x_0)^2 + (y - y_0)^2 - a^2 = 0$. Using complex numbers $z_0 = x_0 + iy_0$ and $z = x + iy$, this can be expressed as

$$|z - z_0|^2 = (z - z_0)(\bar{z} - \bar{z}_0) = a^2. \tag{1.22a}$$

Alternatively, taking c to be a real number $c = |z_0|^2 - a^2$,

$$z\bar{z} - \bar{z}_0 z - z_0 \bar{z} + c = 0 \qquad (c - |z_0|^2 < 0). \tag{1.22b}$$

1.2.2 *Polar representation of complex numbers*

Once introducing the polar coordinate on a two-dimensional plane, the coordinate x and y of a point (x, y) can be expressed as

$$x = r\cos\theta, \qquad y = r\sin\theta, \tag{1.23a}$$

where r is the distance from the origin $(0, 0)$ to the point (x, y)

$$r = \sqrt{x^2 + y^2} = |z| \tag{1.23b}$$

and θ is the angle of rotation counterclockwise from the x axis to the vector (x, y)[6]

$$\theta = \arctan\frac{y}{x} \quad \begin{cases} 0 \leq \theta \leq \pi & (\text{mod } 2\pi) \quad (y \geq 0) \\ -\pi < \theta < 0 & (\text{mod } 2\pi) \quad (y < 0) \end{cases}. \tag{1.23c}$$

The complex number $z = x + iy$ is represented, by using Eq. (1.23a), as Fig. 1.2

$$z = x + iy = r(\cos\theta + i\sin\theta), \tag{1.24}$$

---

[6]For two real numbers a and b, when b is expressed as $b = a + 2\pi n$ with an appropriate integer n, one can write "$a = b \pmod{2\pi}$."

r is the **absolute value** $|z|$ of the complex number z. θ is called the **argument** of the complex number z and is written as $\arg z$,

$$\theta = \arg z. \tag{1.25}$$

However, if it is written as in Eq. (1.23c), the argument has an indefiniteness of an integer multiple of 2π and is not uniquely determined. Therefore, it must be determined in advance that the value of the argument θ has no indefiniteness. $z = 0$ is a complex number whose absolute value is 0 and the argument is indefinite.

The argument may be stated with the specific limitation from $-\pi$ to π (or from 0 to 2π). In that case, it is written as $\text{Arg}\, z$, with the first letter of arg as uppercase rather than lowercase. Thus, when $z \neq 0$,

$$\arg z = \text{Arg}\, z + 2n\pi \qquad (n \text{ is an appropriate integer}),$$
$$-\pi < \text{Arg}\, z \leq \pi. \tag{1.26}$$

$\text{Arg}\, z$ is called the **principal value** of the argument.

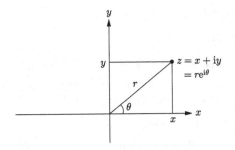

Fig. 1.2   Polar representation of a complex number.

1.2.3   *Euler's formula and the polar form of complex numbers*

Define a function $e^{i\theta}$ of θ, as follows:

$$e^{i\theta} = \cos\theta + i\sin\theta. \tag{1.27}$$

This is called **Euler's formula**. The complex number z can be expressed, with the polar coordinate and Euler's formula, as

$$z = x + iy = r(\cos\theta + i\sin\theta) = re^{i\theta}. \tag{1.28}$$

This is called the **polar form** of a complex number. e is the base of the natural logarithm (or a number called **Napier's number**), a transcendental number $e = 2.7182818284\cdots$.

Remark 1.1. Exponential function is defined generally in Sec. 3.5.1. Here, Eq. (1.27) should be understood as the definition of a function $e^{i\theta}$, whatever a value of e is. The definition and the convergence of the exponential function $e^{i\theta}$ will be discussed, in Examples 1.7 and 1.8, and in Sec. 3.3.1. We will make a simple description here. The exponential function is defined in the form of the following infinite series.

$$e^{i\theta} = \sum_{n=0}^{\infty} \frac{1}{n!}(i\theta)^n.$$

In Example 1.7, we will discuss the "absolute convergence" of this infinite series. If the above series converges, the real parts and imaginary parts can be rearranged, and rewritten using the equation for the Taylor series of a trigonometric function, it produces

$$e^{i\theta} = \sum_{n=0}^{\infty} \frac{1}{n!}(i\theta)^n = \sum_{n=0}^{\infty} \frac{(-1)^n \theta^{2n}}{(2n)!} + i \sum_{n=0}^{\infty} \frac{(-1)^n \theta^{2n+1}}{(2n+1)!}$$

$$= \cos\theta + i\sin\theta.$$

This is Euler's formula.

If the polar form is used, the meaning of the multiplication and division of complex numbers can be geometrically understood. Using the addition theorem for a trigonometric function, one obtains

$$\begin{aligned}
e^{i(\theta_1+\theta_2)} &= \cos(\theta_1+\theta_2) + i\sin(\theta_1+\theta_2) \\
&= (\cos\theta_1\cos\theta_2 - \sin\theta_1\sin\theta_2) + i(\sin\theta_1\cos\theta_2 + \cos\theta_1\sin\theta_2) \\
&= (\cos\theta_1 + i\sin\theta_1)(\cos\theta_2 + i\sin\theta_2) \\
&= e^{i\theta_1}e^{i\theta_2}.
\end{aligned}$$

Therefore, once we write as $z_1 = r_1\exp(i\theta_1)$, $z_2 = r_2\exp(i\theta_2)$, the product is

$$z_1 z_2 = r_1 r_2 \exp\{i(\theta_1+\theta_2)\}. \tag{1.29}$$

Then the absolute value of $z_1 z_2$ is the product of each absolute value, and the argument equals the sum of the arguments:

$$|z_1 z_2| = |z_1||z_2|, \tag{1.30a}$$

$$\arg(z_1 z_2) = \arg z_1 + \arg z_2. \tag{1.30b}$$

The sum of the arguments of Eq. (1.30b) must be treated carefully. Let us consider a simple example. In the case of $z_1 = z_2 = -1 = e^{i\pi}$, $z_1 z_2 = 1$, so one might expect that $\arg(z_1 z_2) = 0$. However, the correct answer is

$\arg z_1 + \arg z_2 = 2\pi$. In other words, since values of z_1 and z_2 are specified by unique absolute values and arguments as above, it is also determined that the absolute value is 1 and the argument is $\pi + \pi = 2\pi$ for $z_1 z_2$ as well. From this, readers can see that as the value is given to a complex number, the only indefiniteness is in $2n\pi$ as the argument. Therefore, for Eq. (1.30b), it means that "both sides are equal as sets". And, for the values of the arguments, both are equal with modulo 2π.

$$\arg(z_1 z_2) = \arg z_1 + \arg z_2 \quad (\mathrm{mod}\ 2\pi), \tag{1.31a}$$

$$\mathrm{Arg}\,(z_1 z_2) = \mathrm{Arg}\,z_1 + \mathrm{Arg}\,z_2 \quad (\mathrm{mod}\ 2\pi). \tag{1.31b}$$

Example 1.3. When the complex number z is multiplied by the imaginary unit $\mathrm{i} = \mathrm{e}^{\mathrm{i}\pi/2}$, the point (x, y) on a complex plane rotates by just $\pi/2$ around the origin (counterclockwise rotation by $\pi/2$). On the other hand, multiplying $-\mathrm{i} = \mathrm{e}^{-\mathrm{i}\pi/2}$ produces rotation by $-\pi/2$ (clockwise rotation by $\pi/2$).

1.2.4 *Exponentiation and nth root of a complex number*

If the polar form is used, it is also simple and easily comprehensible to express the powers of complex numbers.

The polar form $z = r\,\mathrm{e}^{\mathrm{i}\theta}$ is raised to the n-th power, we have, using Eq. (1.29),

$$z^n = r^n(\mathrm{e}^{\mathrm{i}\theta})^n = r^n \mathrm{e}^{\mathrm{i}n\theta}. \tag{1.32}$$

Example 1.4. If both sides of $(\mathrm{e}^{\mathrm{i}\theta})^n = \mathrm{e}^{\mathrm{i}n\theta}$ are expressed using trigonometric functions, **de Moivre's formula** can be derived.

$$(\cos\theta + \mathrm{i}\sin\theta)^n = \cos n\theta + \mathrm{i}\sin n\theta. \tag{1.33}$$

So what happens with the nth root of a complex number? Consider w, which is the nth root of z (where n is a natural number).

$$w = z^{1/n}, \qquad w^n = z. \tag{1.34}$$

When one write $z = r\,\mathrm{e}^{\mathrm{i}\theta}$, $w = \rho\mathrm{e}^{\mathrm{i}\phi}$ in the second equation and compare them, we obtain

$$\rho^n = r, \qquad \mathrm{e}^{\mathrm{i}n\phi} = \mathrm{e}^{\mathrm{i}\theta}.$$

Because $2m\pi$ is the only indefiniteness in the argument of the exponent part of the second equation, $\mathrm{i}n\phi = \mathrm{i}(\theta + 2m\pi)$, so

$$\rho = \sqrt[n]{r}, \qquad \phi = \frac{\theta}{n} + \frac{2m\pi}{n}. \tag{1.35}$$

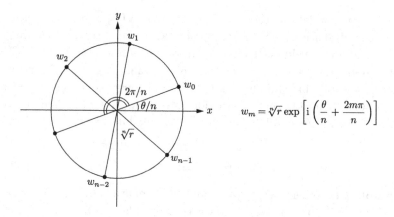

Fig. 1.3   Distribution of power roots.

Here, m is an appropriate integer $(m = 0, 1, 2, 3, \cdots, n - 1)$. Note that when $m \geq n$, two cases

$$\phi = \frac{\theta}{n} + \frac{2m\pi}{n} \quad \text{and} \quad \phi = \frac{\theta}{n} + \frac{2m'\pi}{n} \quad (m' = m - n)$$

result in the same point on the complex w plane. Therefore, the nth roots of z, $w = z^{1/n}$, are distributed on the complex plane at n equally spaced points on a circle of radius $\sqrt[n]{r}$, with its center at the origin (Fig. 1.3). The nth root $w = z^{1/n}$ is a **multi-valued function** (called an n-valued function) by which one point on the complex z plane is mapped to n points on the complex w plane.

Example 1.5. Consider the complex number $i^{1/2}$

$$i^{1/2} = e^{i(\pi/2 + 2n\pi)/2}.$$

So

$$i^{1/2} = \cos\frac{\pi}{4} + i\sin\frac{\pi}{4} \quad (n = 0), \qquad \cos\frac{5\pi}{4} + i\sin\frac{5\pi}{4} \quad (n = 1)$$

and then we obtain

$$i^{1/2} = \frac{1}{\sqrt{2}} + i\frac{1}{\sqrt{2}}, \qquad -\frac{1}{\sqrt{2}} - i\frac{1}{\sqrt{2}}.$$

1.3   Sequence and series of complex numbers

1.3.1   *Sequence and limit*

Definition 1.8 (Limit of a sequence). Consider a **sequence** of complex numbers $\{z_n\}$ $(n = 1, 2, 3, \cdots)$. This is a sequence of points on the complex plane. For this sequence and the complex number c, when

$$\lim_{n\to\infty} |z_n - c| = 0, \tag{1.36}$$

it is said that "sequence $\{z_n\}$ **converges** to a complex number c". This can be written alternatively as

$$\lim_{n\to\infty} z_n = c. \tag{1.37}$$

The complex number c is called the **limit** of the sequence $\{z_n\}$. A sequence that does not converge is said to **diverge**.

Equations (1.36) and (1.37) can be restated in the following way. For any arbitrary positive number ε, there exists a natural number N, such that for all $n > N$,

$$|z_n - c| < \varepsilon. \tag{1.38}$$

This is shown in Fig. 1.4.

From the definition, we get

$$\lim_{n\to\infty} |z_n - c| = 0 \iff \begin{cases} \lim_{n\to\infty} \mathrm{Re}\,(z_n - c) = 0 \\ \lim_{n\to\infty} \mathrm{Im}\,(z_n - c) = 0 \end{cases} \iff \begin{cases} \lim_{n\to\infty} \mathrm{Re}\,z_n = \mathrm{Re}\,c \\ \lim_{n\to\infty} \mathrm{Im}\,z_n = \mathrm{Im}\,c \end{cases}.$$

Thus, the fact that the sequence $\{z_n\}$ of complex numbers converges is equivalent to the fact that two sequences of real numbers $\{x_n\}$ and $\{y_n\}$

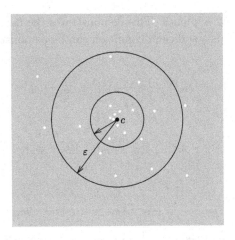

Fig. 1.4   Convergence of a sequence of complex numbers $\{z_n\}$. An infinite number of points exists within a circle of the center c and radius ε, and only a finite number of points exist outside it.

both converge. Therefore, the conditions for judging the convergence of a complex sequence are no different from those for a sequence of real numbers.

Theorem 1.2. (Cauchy's convergence test for a sequence) *If, for any arbitrary positive number ε, there is a positive integer N such that for all natural numbers m, $n > N$*

$$|z_n - z_m| < \varepsilon, \tag{1.39}$$

the sequence $\{z_n\}$ converges. The converse is also true. A sequence as shown in Eq. (1.39), is called **Cauchy sequence,** *or a* **fundamental sequence.**

Proof. The conditions of this theorem are the conditions for the sequence to converge. In other words, if the sequence converges, the fact that it is a Cauchy sequence can be demonstrated as follows: If $\{z_n\}$ converges to c, for any arbitrary positive number $\varepsilon/2$, there exists an appropriate natural number N, and for all $n, m > N$,

$$|z_n - c| < \frac{\varepsilon}{2}, \quad |z_m - c| < \frac{\varepsilon}{2}.$$

Therefore,

$$|z_n - z_m| = |(z_n - c) - (z_m - c)| \le |z_n - c| + |z_m - c| < \varepsilon.$$

Thus, it is a Cauchy sequence. This demonstrates the necessary condition.

The fact that it is a sufficient condition can be demonstrated as follows: Since

$$|z_n - z_m|^2 = |x_n - x_m|^2 + |y_n - y_m|^2,$$

then

$$|z_n - z_m| \ge |x_n - x_m|, \quad |y_n - y_m|.$$

Thus

$$|z_n - z_m| < \varepsilon \Rightarrow |x_n - x_m| < \varepsilon, \quad |y_n - y_m| < \varepsilon.$$

In short, if the complex sequence $\{z_n\}$ is a Cauchy sequence, both sequences $\{x_n\}$ and $\{y_n\}$ are also Cauchy sequences. A sequence of real numbers forms a Cauchy sequence is a necessary and sufficient condition for that sequence to converge. Therefore, sequences $\{x_n\}$ and $\{y_n\}$ converge and, consequently, the sequence z_n converges.[7]

This completes the proof. $\qquad\square$

Theorem 1.3. *When two sequences $\{z_n\}$ and $\{w_n\}$ converge, then*

$$\lim_{n\to\infty}(z_n \pm w_n) = \lim_{n\to\infty} z_n \pm \lim_{n\to\infty} w_n \qquad \text{(double sign corresponds).}$$
$$(1.40)$$

Theorem 1.4. *When two sequences $\{z_n\}$ and $\{w_n\}$ converge, then*

$$\lim_{n\to\infty} z_n w_n = \lim_{n\to\infty} z_n \lim_{m\to\infty} w_m. \qquad (1.41a)$$

In addition to that, if the limit of w_m is not 0, then

$$\lim_{n\to\infty} \frac{z_n}{w_n} = \frac{\displaystyle\lim_{n\to\infty} z_n}{\displaystyle\lim_{m\to\infty} w_m}. \qquad (1.41b)$$

Proving these formulae is not difficult, so let us leave it to readers.

1.3.2 *Series and its convergence*

Let us consider an **infinite series** of complex numbers, and the **convergence** of it. Later, when a function of a complex variable z is defined as an infinite series (a power series of z) or their properties are discussed, they are often discussed on the basis of the properties of a power series.

---

[7]We will present $|x_n - x_m| < \varepsilon \Rightarrow \lim x_n = a$ to remember the core of the proof concerning a sequence of real numbers.

First, it should be shown that x_n is bounded. When $n > N$, for any arbitrary integer ε,

$$|x_n - x_N| < \varepsilon, \qquad x_N - \varepsilon < x_n < x_N + \varepsilon.$$

Therefore, for $n > N$, x_n is bounded. Therefore, a finite number of terms x_m ($m \le N$) are added and $\{x_n\}$ is bounded.

Now, letting the upper and lower limit of $x_n, x_{n+1}, x_{n+2}, \cdots$ for any arbitrary n, be u_n, l_n,

$$l_1 \le l_2 \le \cdots \le l_n \le \cdots \le u_n \le \cdots \le u_2 \le u_1.$$

$\{l_n\}, \{u_n\}$ are each bounded monotone sequences, so they converge. Furthermore,

$$u_n - l_n \le \varepsilon$$

and ε can be made smaller to any desired extent by making n sufficiently large. From the above, it follows that l_n, u_n converge, and that a exists for which $l_n \to a$, $u_n \to a$. This demonstrates that

$$\lim_{m\to\infty} x_n = a.$$

This discussion is called the **method of nested intervals**.

Definition 1.9 (Series and its convergence).

$$\sum_{m=1}^{\infty} z_m = z_1 + z_2 + z_3 + \cdots \tag{1.42}$$

is called a series (of complex numbers). The partial sum of a complex series is defined as

$$S_N = \sum_{m=1}^{N} z_m. \tag{1.43}$$

When the sequence $\{S_N\}$ converges on S, it is said that "series $\sum_{n=1}^{\infty} z_n$ **converges** on S". When the sequence $\{S_N\}$ diverges, it is said that "series $\sum_{n=1}^{\infty} z_n$ **diverges**".

Saying that the series $\sum_{n=1}^{\infty} z_n$ converges is equivalent to saying that the two series of real numbers $\sum_{n=1}^{\infty} x_n$, $\sum_{n=1}^{\infty} y_n$ both converge.

Theorem 1.5. (Cauchy's convergence test for a partial sum) *If a series $\sum_{n=1}^{\infty} z_n$ converges, for any positive number ε there is a fixed integer N such that*

$$|z_{n+1} + z_{n+2} + \cdots + z_{n+p}| < \varepsilon \qquad (n > N, \; p > 0). \tag{1.44}$$

The converse is also true.

As $z_{n+1} + z_{n+2} + \cdots + z_{n+p} = S_{n+p} - S_n$, this theorem is an alternative statement of the convergence test Th. 1.2 for a sequence $\{S_n\}$. We will leave the proof to readers.

Theorem 1.6. *If $\sum_{n=1}^{\infty} z_n$ converges, $\lim_{n \to \infty} z_n = 0$.*

Proof. One should set $p = 1$ in the theorem of convergence test Th. 1.5. $\square$

Definition 1.10. When series $\sum_{n=1}^{\infty} |z_n|$ converges, it is said that "the complex series $\sum_{n=1}^{\infty} z_n$ **converges absolutely**", and the series $\sum_{n=1}^{\infty} z_n$ is called an **absolutely convergent series**.

Theorem 1.7. *If series* $\sum\limits_{n=1}^{\infty} z_n$ *converges absolutely, this series converges.*

Proof. For a set of complex numbers $z_{n+1}, z_{n+2}, \cdots, z_{n+p}$,

$$|z_{n+1} + z_{n+2} + \cdots + z_{n+p}| \leq |z_{n+1}| + |z_{n+2}| + \cdots + |z_{n+p}|.$$

If series $\sum\limits_{n=1}^{\infty} z_n$ absolutely converges,

$$|z_{n+1}| + |z_{n+2}| + \cdots + |z_{n+p}| < \varepsilon.$$

Therefore,

$$|z_{n+1} + z_{n+2} + \cdots + z_{n+p}| < \varepsilon.$$

$\square$

Several theorems concerning absolutely convergent series are stated below. Readers should prove them, as none of them are difficult.

Theorem 1.8. *An absolutely convergent series remains an absolutely convergent series even if the order of the terms is changed, and the value is the same as the sum of the original series.*

Theorem 1.9. *A series created as a single term by combining several subsequent terms in an absolutely convergent series is also absolutely convergent, and its value is the same as the sum of the original series.*

Theorem 1.10. *If* $\sum z_n$ *and* $\sum w_n$ *converge absolutely,* $\sum(z_n + w_n)$ *also does so, and is* $\sum(z_n + w_n) = \sum z_n + \sum w_n$.

Theorem 1.11. *If a sequence of non-negative real numbers* $\{M_n\}$ *exists in which* $|z_n| \leq M_n$ *and a series* $\sum M_n$ *converges,* $\sum z_n$ *converges absolutely.*

Example 1.6. If $|z| < 1$,

$$\sum_{n=1}^{\infty} z^n$$

converges absolutely. Consider a sequence of real numbers $\{M_n\}$ for which $M_n = x^n$, $|z| = x < 1$ (the geometric sequence x^n). Then, the series $\sum z^n$ ($|z| < 1$) converges absolutely.

Example 1.7. A series

$$\sum_{n=0}^{\infty} \frac{z^n}{n!}$$

converges absolutely. That is the case because a series of real numbers $e^x = \sum_{n=0}^{\infty} x^n/n!$, for which $M_n = |z|^n/n!$, $|z| = x$, converges absolutely. This is written as e^z or $\exp z$.

Theorem 1.12. *When $\sum z_n$ and $\sum w_n$ converge absolutely, take all combinations $z_n w_m$ of each term. The series $\sum z_n w_m$, with the terms rearranged into any arbitrary order, converges absolutely, and their sum is equal to the product of the sums of each series.*

$$\sum_{n,m=1}^{\infty} z_n w_m = \sum_{n=1}^{\infty} z_n \sum_{m=1}^{\infty} w_m. \tag{1.45}$$

Proof. When a partial sum of the series $\sum z_n w_m$ is selected, let the maximum values of the indices n, m in the partial sum, be K and L. In this case, the partial sum is evaluated as

$$\sum |z_n w_m| \leq \sum_{n=1}^{K} |z_n| \sum_{m=1}^{L} |w_m| \leq \sum_{n=1}^{\infty} |z_n| \sum_{m=1}^{\infty} |w_m|.$$

As $\sum |z_n|$, $\sum |w_m|$ converge absolutely, the left-hand side is finite, regardless of the method used to select the subsequence. Therefore, the sum $\sum z_n w_m$ of the infinite series converges absolutely. This does not depend on the order of $z_n w_m$. From the above, an appropriate reordering of the sequence of terms in $\sum_{n,m=1}^{\infty} z_n w_m$ produces

$$\sum_{n,m=1}^{\infty} z_n w_m = \sum_{n=1}^{\infty} z_n \sum_{m=1}^{\infty} w_m.$$

$\square$

Example 1.8. We define

$$e^z = \sum_{n=1}^{\infty} \frac{z^n}{n!} \tag{1.46}$$

and then, it is possible, with care on

$$\frac{(z_1 + z_2)^n}{n!} = \sum_{\substack{l+m=n \\ l \geq 1, m \geq 1}} \frac{z_1^l z_2^m}{l! m!}$$

to immediately derive

$$e^{z_1 + z_2} = e^{z_1} e^{z_2} \tag{1.47}$$

from Th. 1.12.

Chapter 2

Complex Functions and Holomorphy

We will discuss, in this chapter, complex analysis, *i.e.* differentiability, which is one of the basic properties of complex functions, and the properties derived directly from it. Because the complex number z is a point on a complex plane, there are an infinite number of paths on the complex plane to make z approach z_0 to define the differentiation. The fact that a constant value exists independently of how this limit is approached is inseparable from the definition of the differentiation of a complex function.

2.1 Complex functions and their continuity

Definition 2.1 (Complex function). When a domain on the complex w plane is mapped from a domain of the complex z plane, and written as

$$w = f(z), \tag{2.1}$$

$f(z)$ is called a function of a complex variable z, or a **complex function**.

For a complex function $w = f(z)$, we assume that w approaches w_0 as much as one wants, when z approaches z_0 close enough. This means that for any positive number ε, there is a positive number δ such that

$$|f(z) - w_0| < \varepsilon, \quad \text{whenever} \quad |z - z_0| < \delta. \tag{2.2}$$

This can be also written as

$$\lim_{z \to z_0} f(z) = w_0. \tag{2.3}$$

w_0 is called "the limit of $f(z)$ as z approaches z_0".

Definition 2.2 (Continuity). It is said that "$f(z)$ is continuous at $z = z_0$" (Fig. 2.1), when $\lim_{z \to z_0} f(z) = f(z_0)$, *i.e.* an appropriate positive number δ exists for any given positive number ε, such that

$$|f(z) - f(z_0)| < \varepsilon \text{ for all } z \text{ of } |z - z_0| < \delta. \tag{2.4}$$

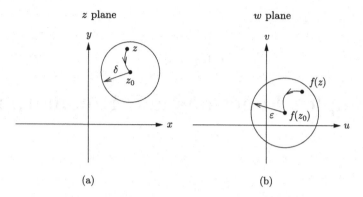

Fig. 2.1 Continuity of functions: The mapping $w = f(z)$ of a point z within a circle of radius δ centered at z_0 is within a circle of radius ε centered at $w_0 = f(z_0)$.

In general, mapping from **two-dimensional vector** space (x, y) to two-dimensional vector space (u, v) can be expressed as

$$f(z) = u(x, y) + iv(x, y).$$

Furthermore, as $x = (z + \bar{z})/2$ and $y = (z - \bar{z})/(2i)$, it can also be expressed as

$$f = u(z, \bar{z}) + iv(z, \bar{z}).$$

The target in our complex analysis is not functions that cannot be expressed without using $\bar{z}$, but functions expressed as

$$w = f(z) = u(z) + iv(z), \tag{2.5}$$

such as $w_1 = z^2$. On the other hand, a function $w_2 = x^2 + y^2$ is expressed as $w_2 = z\bar{z} = |z|^2$, and it is not possible to write this function w_2 with z alone, without using $\bar{z}$ or $|z|$. This kind of function is not a central issue in this book (see Example 2.7).

If $\lim_{z \to z_0} f(z)$ and $\lim_{z \to z_0} g(z)$ exist for the complex functions $f(z)$ and $g(z)$, the following properties hold.

(1) $\lim_{z \to z_0} [f(z) + g(z)] = \lim_{z \to z_0} f(z) + \lim_{z \to z_0} g(z)$,

(2) $\lim_{z \to z_0} cf(z) = c \lim_{z \to z_0} f(z)$ (c is an arbitrary complex number).

(3) $\lim_{z \to z_0} f(z) g(z) = \lim_{z \to z_0} f(z) \cdot \lim_{z \to z_0} g(z)$,

(4) $\lim_{z \to z_0} \dfrac{f(z)}{g(z)} = \dfrac{\lim_{z \to z_0} f(z)}{\lim_{z \to z_0} g(z)}$, where $\lim_{z \to z_0} g(z) \neq 0$.

Readers could prove them easily.

2.2 The differentiability and holomorphy of complex functions

2.2.1 *Differentiation of complex functions*

Definition 2.3 (ε-neighborhood, neighborhood).
The ε-neighborhood of z_0 is a set of all points on Gauss plane whose distance to z_0 is less than $\varepsilon(> 0)$,

$$\{z \mid |z - z_0| < \varepsilon\}. \tag{2.6}$$

A set V on Gauss plane is called a "neighborhood of a point z_0" if an ε-neighborhood of z_0 is contained in V.

Definition 2.4 (Differentiation). For a complex function $w = f(z)$ that is defined in a domain D on a complex z plane, if

$$\lim_{z \to z_0} \frac{f(z) - f(z_0)}{z - z_0} \tag{2.7}$$

exists uniquely (finitely determined) for any given point z_0 within D, we say that "$f(z)$ is **differentiable** in domain D". This Eq. (2.7) is called a **differential coefficient** or **derivative**, and is written as $f'(z_0)$ or $\dfrac{df}{dz}\Big|_{z=z_0}$.

$$f'(z_0) = \frac{df}{dz}\Big|_{z=z_0} = \lim_{z \to z_0} \frac{f(z) - f(z_0)}{z - z_0}. \tag{2.8}$$

Here this is expressed simply as $\lim_{z \to z_0}$ (bring z closer to z_0 to any arbitrary extent), but how to approach (the path of approach on the two-dimensional plane) is not specified. In other words, there must exist a unique limit value (2.8) that does not depend on the path of whatever approach is used. Fruitful properties of complex functions can be derived from this condition.

Various subsequent discussions can be facilitated by writing the condition of differentiability as follows:

$$\Delta f \equiv f(z + \Delta z) - f(z) = f'(z)\Delta z + \delta \Delta z. \tag{2.9}$$
$$\text{where } \delta \to 0 \text{ when } \Delta z \to 0.$$

Definition 2.5 (Holomorphy, regularity). A function that is differentiable for any given point in domain D is called "**holomorphic (regular)** in domain D". Saying that the complex function $f(z)$ is "holomorphic at the point z_0" means that "it is differentiable at the point z_0", and the point z_0 is called a **holomorphic point (regular point)** of $f(z)$. A function that is holomorphic (in domain D) is called a **holomorphic function**. When a function $f(z)$ is "not holomorphic" ("not regular" or "singular") at $z = z_0$, the point z_0 is called a **singularity**.

2.2.2   *Differential formulae*

The basic formulae related to differentiation, such as the differentiation of powers and nth roots, are derived directly from the definition of differentiation.

Example 2.1. Taking n to be a positive integer, the differentiation of the power z^n is, from the formula for binomial expansion

$$(z + \Delta z)^n = z^n + nz^{n-1}\Delta z + \frac{1}{2}n(n-1)z^{n-2}(\Delta z)^2 + \cdots,$$

we get a formula

$$\frac{dz^n}{dz} = \lim_{\Delta z \to 0} \frac{(z + \Delta z)^n - z^n}{\Delta z} = nz^{n-1}. \tag{2.10}$$

Example 2.2. Taking n to be a positive integer, and assuming the nth root $z^{1/n}$ be $(z + \Delta z)^{1/n} = a$, $z^{1/n} = b$, the equation

$$\frac{(z + \Delta z)^{1/n} - z^{1/n}}{\Delta z} = \frac{a - b}{a^n - b^n}$$

$$= \frac{a - b}{(a - b)(a^{n-1} + a^{n-2}b + a^{n-3}b^2 + \cdots + ab^{n-2} + b^{n-1})}$$

leads to

$$\frac{dz^{1/n}}{dz} = \lim_{a \to b} \frac{1}{a^{n-1} + a^{n-2}b + a^{n-3}b^2 + \cdots + ab^{n-2} + b^{n-1}}$$

$$= \frac{1}{n}z^{(1/n)-1}. \tag{2.11}$$

Furthermore, $z^{1/n}$ cannot be differentiated when $z = 0$.

We will now state the various formulae for the differentiation of complex functions:

$$\frac{d}{dz}[cf(z)] = c\frac{df(z)}{dz}, \tag{2.12a}$$

$$\frac{d}{dz}[f(z) \pm g(z)] = \frac{df(z)}{dz} \pm \frac{dg(z)}{dz}, \tag{2.12b}$$

$$\frac{d}{dz}(f(z)\,g(z)) = \frac{df(z)}{dz}g(z) + f(z)\frac{dg(z)}{dz}, \tag{2.12c}$$

$$\frac{d}{dz}\left[\frac{f(z)}{g(z)}\right] = \frac{\frac{df(z)}{dz}g(z) - f(z)\frac{dg(z)}{dz}}{g(z)^2}. \tag{2.12d}$$

These proofs are formally identical to those for real functions and are simple when Eq. (2.9) is employed.

Theorem 2.1 (Differentiation of composite function). *When $g(z)$ is differentiable at $z = z_0$, and $f(w)$ is differentiable at $w = w_0 = g(z_0)$, $f(g(z))$ is differentiable at $z = z_0$ and*

$$\frac{df(g(z_0))}{dz} = \frac{df(w(z_0))}{dw} \cdot \frac{dg(z_0)}{dz} \qquad (2.13)$$

is true.

Proof. If the variation in f corresponding to a variation Δg in g is written as Δf, differentiability produces

$$\Delta f = f'(g)\Delta g + \gamma \Delta g, \qquad \Delta g = g'(z)\Delta z + \gamma' \Delta z.$$

In this situation, $\gamma \to 0$ when $\Delta g \to 0$, and $\gamma' \to 0$ when $\Delta z \to 0$. Therefore,

$$\frac{\Delta f}{\Delta z} = (f'(g) + \gamma)(g'(z) + \gamma') = f'(g)g'(z) + (f'(g)\gamma' + \gamma g'(z) + \gamma \gamma').$$

When $\Delta z \to 0$, it follows that $\gamma \to 0$, $\gamma' \to 0$. Then the parenthetical expression at the end of the right-hand side converges to 0, and Eq. (2.13) is derived. $\qquad \square$

Example 2.3. Taking n to be a positive integer, consider the differentiation of the inverse power z^{-n}. If we take $f(w) = w^n$, $w(z) = z^{-1}$ in the differential formula (2.13) for a composite function, it follows that

$$\frac{d}{dz}(z^{-n}) = \frac{dw^n}{dw} \cdot \frac{d}{dz}\left(\frac{1}{z}\right) = n\left(\frac{1}{z}\right)^{n-1} \cdot \left(-\frac{1}{z^2}\right) = -nz^{-n-1}. \qquad (2.14)$$

Example 2.4. Taking q to be the rational number $q = m/n$ (where n is a positive integer), consider the differentiation of z^q. Equation (2.14) applies here, so it makes no difference whether m is a positive or a negative integer. In Eq. (2.13), $f(w) = w^m$ and $w(z) = z^{1/n}$ can be assumed.

$$(z^{m/n})' = \frac{dw^m}{dw} \cdot \frac{dz^{1/n}}{dz} = m(z^{1/n})^{m-1} \cdot \frac{1}{n}z^{1/n-1} = \frac{m}{n}z^{m/n-1} = qz^{q-1}.$$

From the above, we have derived the following for the power z^q, when q is a rational number:

$$\frac{dz^q}{dz} = qz^{q-1}. \qquad (2.15)$$

Consider further z^q, where q is the rational number $q = m/n$. When the argument θ varies from 0 to 2π in $z = \rho e^{i\theta}$, the function $z^{m/n}$ is not continuously connected from the value at $\theta = 2\pi$ to the value at $\theta = 0$ (Fig. 2.2). That is because $z^{m/n}$ is not single-valued on the complex z

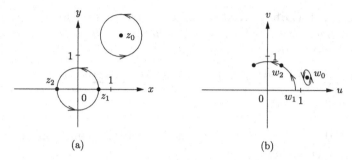

Fig. 2.2   Behavior of $w = z^{m/n}$. z_0 is mapped to w_0, and z_1 and z_2 are mapped to w_1 and w_2, respectively.

plane. Furthermore, this means that in the neighborhood of $z = 0$, even if the value $z_1^{m/n}$ at $z = z_1$ is uniquely determined, including the argument, the value $z_2^{m/n}$ at the point z_2, which is very slightly displaced from z_1, is sometimes not uniquely determined. That is the case because if the path of the movement from z_1 to z_2 passes through $z = 0$, the argument of z becomes indeterminate. This situation which occurs with $z^{m/n}$ is only limited to the neighborhood of the origin $z = 0$ for z^q. As long as the point z moves at a finite distance from the origin, the uniqueness of the values is never destroyed. From the above, it is clear that the function $z^q = z^{m/n}$ is not holomorphic at $z = 0$.[1]

From the above example, it can be said that "The formula of differentiation for a complex function is the same as that for a real function".

2.2.3   *Cauchy–Riemann relationship and derivatives of inverse functions*

A holomorphic function in a given domain is a function that is differentiable in the domain. When it is a real function, differentiability may not seem to be a very stringent condition, but the independence of the path $z \to z_0$ for the derivative of a complex function is an extraordinarily restricted condition, and many basic properties are derived from that unique condition.

---

[1]Beyond the rational power $z^{m/n}$, the same is true for irrational powers. But irrational powers have not yet been defined here. We will examine the details later, and this kind of point $z = 0$ is called a branch point.

2.2.3.1   *Cauchy–Riemann relation*

Cauchy–Riemann relation below provides the one necessary and sufficient condition for a holomorphic function. It shows how stringent the condition is for a holomorphic function.

Theorem 2.2 (Cauchy–Riemann relation). *When the real and imaginary parts of $f = u + iv$ are one-time continuously differentiable (the first derivatives exist and are continuous) for x and y in the neighborhood of $z = z_0$, the necessary and sufficient condition for $f(z) = u + iv$ to be holomorphic is that the following must be satisfied:*

$$\frac{\partial u}{\partial x} = \frac{\partial v}{\partial y}, \qquad \frac{\partial u}{\partial y} = -\frac{\partial v}{\partial x}. \tag{2.16}$$

Equation (2.16) is called **Cauchy–Riemann relation**.

Proof. The proof of Cauchy–Riemann relation is simple. It is basically that the result must be independent of approaching path $z \to z_0$. So taking $\Delta z = \Delta x + i\Delta y$ to be the case when $z + \Delta z$ approaches z, consider the differentiation of the following two paths:

(1) $\Delta y = 0$, $\Delta x \to 0$,
(2) $\Delta x = 0$, $\Delta y \to 0$.

Writing (1) and (2),

$$f'(z) = \lim_{\Delta x \to 0} \left[\frac{u(x + \Delta x, y) - u(x, y)}{\Delta x} + i\frac{v(x + \Delta x, y) - v(x, y)}{\Delta x} \right]$$

$$= \frac{\partial u}{\partial x} + i\frac{\partial v}{\partial x}, \tag{2.17a}$$

$$f'(z) = \lim_{\Delta y \to 0} \left[\frac{u(x, y + \Delta y) - u(x, y)}{i\Delta y} + i\frac{v(x, y + \Delta y) - v(x, y)}{i\Delta y} \right]$$

$$= -i\frac{\partial u}{\partial y} + \frac{\partial v}{\partial y} \tag{2.17b}$$

are derived. Comparing the real and imaginary parts of the final terms of these equations produces

$$\frac{\partial u}{\partial x} = \frac{\partial v}{\partial y}, \qquad \frac{\partial u}{\partial y} = -\frac{\partial v}{\partial x}.$$

This indicates that Eq. (2.16) is a necessary condition.

Next, consider the sufficient condition. Assume partial differential coefficients u_x, u_y, v_x, v_y to be continuous and to satisfy Eq. (2.16). In that case, from the definition of partial differentiation, one gets

$$u = u_0 + a\Delta x + b\Delta y + \varepsilon\sqrt{(\Delta x)^2 + (\Delta y)^2},$$
$$v = v_0 - b\Delta x + a\Delta y + \varepsilon'\sqrt{(\Delta x)^2 + (\Delta y)^2}.$$

Both ε and ε' become 0 when $|\Delta z| \to 0$. From $f = u + iv$,

$$f = f_0 + (a - ib)\Delta z + \varepsilon''\sqrt{(\Delta x)^2 + (\Delta y)^2}.$$

Here, $\varepsilon'' \to 0$, when $|\Delta z| \to 0$. $f_0 = u_0 + iv_0$ is the value of f at z_0. It follows that

$$f' = a - ib$$

and the differential coefficient exists. This indicates that Eq. (2.16) is a sufficient condition. $\qquad\square$

The above condition is a little too strict. The condition can be relaxed slightly to "the necessary and sufficient condition for $f = u + iv$ to be holomorphic is that u, v must be **totally differentiable** (and therefore partially differentiable), and that Cauchy–Riemann relation must hold".[2] The version with the more strict condition is easier to prove and poses no practical impediment, so we adopt it here.

Let us examine some simple examples.

Example 2.5. Let us consider a holomorphic function $f(z) = z^2$, which is differentiable at any point of a finite distance from the origin on a complex plane. The real and imaginary parts of $f(z)$ are $u(x, y) = x^2 - y^2$ and $v(x, y) = 2xy$, respectively. These u, v satisfy Cauchy–Riemann relation (2.16).

As shown by Goursat theorem in Sec. 7.2, a holomorphic function is infinite-times continuously differentiable. Thus, the real part u and the imaginary part v of the holomorphic function are infinite-times continuously differentiable for x and y. We are not ready to prove this and will accept it as it stands. Differentiating Cauchy–Riemann relation (2.16) once more for x and y produces

$$\frac{\partial^2 u}{\partial x^2} = \frac{\partial^2 v}{\partial x \partial y}, \qquad \frac{\partial^2 u}{\partial y^2} = -\frac{\partial^2 v}{\partial x \partial y}.$$

---

[2]Saying that $u(x, y)$ is totally differentiable for (x, y) means that it can be written as $\Delta u \equiv u(x+\Delta x, y+\Delta y) - u(x, y) = a\Delta x + b\Delta y + \varepsilon\sqrt{(\Delta x)^2 + (\Delta y)^2}$, that a, b is unrelated to Δx, Δy, and that $\varepsilon \to 0$ when $(\Delta x)^2 + (\Delta y)^2 \to 0$.

Then we obtain

$$\frac{\partial^2 u}{\partial x^2} + \frac{\partial^2 u}{\partial y^2} = 0. \tag{2.18a}$$

Similarly, for $v(x, y)$

$$\frac{\partial^2 v}{\partial x^2} + \frac{\partial^2 v}{\partial y^2} = 0. \tag{2.18b}$$

These two equations (2.18a) and (2.18b) are **Laplace equations** in two-dimensions. In general, the solutions of Laplace equation in n-dimensions are called **harmonic functions**.

According to Cauchy–Riemann relation, if either the real part or the imaginary part of a holomorphic function is determined, the other can be determined except a constant term. From the viewpoint of a harmonic function in a two-dimensional space, if one harmonic function $u(x, y)$ is determined, the harmonic function $v(x, y)$ that pairs with it to produce the holomorphic function $f(z) = u + iv$ is determined by Cauchy–Riemann relation. u and v may be called **conjugate** harmonic functions of the other.

Example 2.6. Suppose the real part of the holomorphic function $f(z) = u + iv$ is $u(x, y) = x^2 - y^2$.

$$\Delta u = \left(\frac{\partial^2}{\partial x^2} + \frac{\partial^2}{\partial y^2} \right) u = 2 - 2 = 0.$$

Therefore, u is confirmed to be a harmonic function. From Cauchy–Riemann relation, the imaginary part v must satisfy

$$\frac{\partial v}{\partial x} = -\frac{\partial u}{\partial y} = 2y, \qquad \frac{\partial v}{\partial y} = \frac{\partial u}{\partial x} = 2x.$$

This can be integrated to produce

$$v(x, y) = \int^x dx(2y) = 2xy + \phi(y) = \int^y dy(2x) = 2xy + \psi(x).$$

From these it follows that $\phi(y) = \psi(x) = a$ (constant). Therefore, except a constant term, one obtains $f(z) = (x^2 - y^2) + i2xy = z^2$.

2.2.3.2 *Derivative of inverse function*

The following **derivative formula of inverse function** is derived from Cauchy–Riemann relation and the differential formula for a composite function.

Theorem 2.3 (Derivative of inverse function). *Suppose $f(z)$ is holomorphic at $z = z_0$, and that $f'(z_0) \neq 0$. In that case, $z = g(w)$, which is*

Complex Function Theory

the inverse function of $w = f(z)$, exists in the neighborhood of $z = z_0$. For the differentiation of inverse functions,

$$\frac{\mathrm{d}f}{\mathrm{d}z} = \frac{1}{\left(\dfrac{\mathrm{d}g}{\mathrm{d}w}\right)} \tag{2.19}$$

is obtained.

Proof. The proof is performed in two stages. The first is proof of the existence of an inverse function, and the second is the application of its differential formula. Consider the function $f = u + iv$ of two variables x, y

$$u = u(x, y), \qquad v = v(x, y). \tag{2.20}$$

When Jacobian determinant

$$\frac{D(u, v)}{D(x, y)} = \begin{vmatrix} \dfrac{\partial u}{\partial x} & \dfrac{\partial u}{\partial y} \\ \dfrac{\partial v}{\partial x} & \dfrac{\partial v}{\partial y} \end{vmatrix} = J \tag{2.21}$$

is not zero ($J \neq 0$), the inverse mapping of the above function

$$x = x(u, v), \qquad y = y(u, v) \tag{2.22}$$

is uniquely determined. Therefore the inverse function of the holomorphic function $f(z)$ exists. Jacobian determinant, using Cauchy–Riemann relation, is

$$J = \begin{vmatrix} \dfrac{\partial u}{\partial x} & \dfrac{\partial u}{\partial y} \\ \dfrac{\partial v}{\partial x} & \dfrac{\partial v}{\partial y} \end{vmatrix} = u_x v_y - u_y v_x = u_x{}^2 + v_x{}^2 = |f'(z)|^2,$$

so the necessary and sufficient condition is $f'(z) \neq 0$. This ends the first stage.

The second stage is that for the relation between $f(z)$ and the inverse function $g(w)$

$$w = f(g(w)),$$

applying the differential formula for a composite function, produces

$$1 = \frac{\mathrm{d}w}{\mathrm{d}w} = \frac{\mathrm{d}f}{\mathrm{d}z}\frac{\mathrm{d}g}{\mathrm{d}w}, \tag{2.23}$$

and the target equation (2.19) is obtained. $\qquad\qquad\square$

2.2.4 *Partial differentiation by z and partial differentiation by $\bar{z}$*

We have defined a differentiation by the complex number z, and discussed the properties that are derived therefrom. We now take a slightly different perspective.

Consider the independent variables x, y, and introduce z and $\bar{z}$ by the following variable transformation.

$$z = x + iy, \tag{2.24a}$$

$$\bar{z} = x - iy. \tag{2.24b}$$

There are various convenient benefits to considering z and $\bar{z}$ as independent variables in this way. Conversely, for Eq. (2.24)

$$x = \frac{1}{2}(z + \bar{z}), \tag{2.25a}$$

$$y = \frac{1}{2i}(z - \bar{z}) \tag{2.25b}$$

so the "partial" differentiation by z and $\bar{z}$ may be

$$\frac{\partial}{\partial z} = \frac{1}{2}\left(\frac{\partial}{\partial x} - i\frac{\partial}{\partial y}\right), \tag{2.26a}$$

$$\frac{\partial}{\partial \bar{z}} = \frac{1}{2}\left(\frac{\partial}{\partial x} + i\frac{\partial}{\partial y}\right). \tag{2.26b}$$

Readers should note that the "partial" differentiation used here is not a partial differentiation in the true sense, but rather, the differentiation of the right-hand side of Eq. (2.26) is the definition of the left-hand side (**Wirtinger derivative**).

In Sec. 2.1, we stated that "functions which cannot be expressed without using $\bar{z}$ are not central to our discussion". Let us consider what that means.

Example 2.7. Two statements that $f(z)$ is holomorphic in the domain D, and that $\frac{\partial}{\partial \bar{z}}f = 0$ are equivalent.

Let us demonstrate that. If we write $f = u + iv$, an equation $\frac{\partial}{\partial \bar{z}}f = 0$ can be rewritten as

$$\begin{aligned}
\frac{\partial}{\partial \bar{z}}f &= \frac{1}{2}\left(\frac{\partial}{\partial x} + i\frac{\partial}{\partial y}\right)(u + iv) \\
&= \frac{1}{2}\left(\frac{\partial u}{\partial x} - \frac{\partial v}{\partial y}\right) + \frac{i}{2}\left(\frac{\partial u}{\partial y} + \frac{\partial v}{\partial x}\right) \\
&= 0. \tag{2.27}
\end{aligned}$$

Looking at the real part and the imaginary part here, this is none other than Cauchy–Riemann relation. If we trace this equation back, assuming Cauchy–Riemann relation, we obtain $\dfrac{\partial}{\partial \bar{z}} f = 0$. Thus, we have demonstrated that $\dfrac{\partial}{\partial \bar{z}} f = 0$ holds for a holomorphic function. The above is the meaning of the statement that "functions which cannot be expressed without using $\bar{z}$ are not central to our discussion".

Chapter 3

Elementary Functions

This chapter discusses power series and a few elementary functions. First, define a point of an infinite distance from the origin on a complex plane ("a point at infinity"). This enables us to make a very general discussion of the convergence of power series. The region of convergence of a power series introduces the concept of "convergence radius". This chapter will also organize and discuss an exponential function, trigonometric functions, a logarithmic function, and a general exponential function. To understand the properties of logarithmic functions and general exponential functions, particularly multi-valuedness, we will introduce the concept of "Riemann surfaces".

3.1 Point at infinity

When we have considered points on the complex z plane up to now, we have always set them at a finite distance from the origin. Now, let us consider mapping $w(z) = 1/z$. Points of $z \neq 0$ present no problems, but the point $z = 0$ requires special handling.

When we considered $w = 1/z$, there is a one-to-one correspondence between all points on the complex z plane, except the point $z = 0$, and points on the complex w plane, except the point $w = 0$. It is convenient to generalize this, and define a point to which $z = 0$ should be mapped by $w = 1/z$, or define a point on the z plane that should be mapped to $w = 0$, so that all the points on the complex z plane and the points on the complex w plane are in one-to-one correspondence. The complex number $z = 0$ has an indefinite argument with the absolute value 0. Therefore, the point corresponding to $z = 0$ by $w = 1/z$ should be regarded as having an infinitely large absolute value and an indefinite argument.

Definition 3.1 (Point at infinity). Introduce the point mapped by

$$\lim_{z \to 0} \frac{1}{z}. \tag{3.1}$$

This new point is called a **point at infinity**, and has ∞ as its symbol. The point at infinity has an infinite absolute value and an indefinite argument.

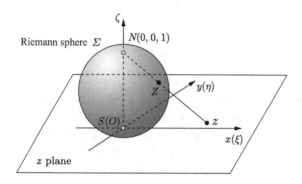

Fig. 3.1   Riemann sphere and the point at infinity.

Let us consider that in more concrete terms. As Fig. 3.1 shows, in three-dimensional space (ξ, η, ζ), the ξ–η plane is regarded to be identical to the complex z plane (x, y). Place a spherical surface Σ with radius $1/2$ such that the point S on the spherical surface contacts the origin O of the ξ–η plane, and let the point symmetrical about the center of the surface from S be $N(0, 0, 1)$. If Z is the point where the line connecting N and z on the plane intersects with the spherical surface Σ, then there is a one-to-one correspondence between point z on the z plane and the point Z on the spherical surface, except the case $Z = N$. If one more point z is added to the complex z plane and made to correspond to N, it makes the complex z plane and the spherical surface Σ fully correspond one-to-one. The new point added to the complex z plane (corresponding to N) is the **point at infinity**.

As has been described, drawing a line joining a point z on the complex z plane and N and extending the point z away from the origin O brings the point Z on Σ closer to N on Σ. Therefore, the point corresponding to z taken to an infinite distance from O can be called the point at infinity. Whichever direction the point z moves in away from the origin, the corresponding point Z approaches N, so the destination to which z is moved infinitely far from the origin can be regarded as a single point, regardless of the direction of separation. Thus, the argument is indefinite. This means that "the point at infinity is a point of an infinite distance from the origin, with an indefinite argument". The spherical surface Σ is called **Riemann sphere**, and the mapping that maps the complex plane onto the Riemann sphere is called a "stereographic projection".

For a point z on the complex z plane, the domain $|z - z_0| < \varepsilon$ for a positive number ε has been called the "ε neighborhood of point z". Corresponding to this, the domain $|z| > R$ on the complex z plane for any arbitrary positive number R is called the **neighborhood of the point at infinity**.

3.2 Power series

3.2.1 *Convergence of a power series*

Saying that a sequence of functions $\{f_n(z)\}_{n=1}^{\infty}$ defined in the domain D "converges" to $f(z)$ in the domain means that, provided a natural number N is chosen appropriately for any arbitrary positive number ε, $|f_n(z) - f(z)| < \varepsilon$ is true for all natural numbers $n \geq N$. When this is true for the same ε and the same N for all values of z in D, it is said to be **uniformly convergent**.

As for the definition of convergence and uniform convergence of an infinite series $G(z) = g_1(z) + g_2(z) + \cdots + g_n(z) + \cdots$, this is equivalent to that of functions $\{s_n(z)\}_{n=1}^{\infty}$ where $s_n(z) = g_1(z) + g_2(z) + \cdots + g_n(z)$.

Series

$$f(z) = a_0 + a_1(z - z_0) + a_2(z - z_0)^2 + \cdots + a_n(z - z_0)^n + \cdots \quad (3.2)$$

is called **power series** around z_0. To avoid a complicated discussion below, we will consider the case of $z_0 = 0$, which is to say, the power series around the origin. This discussion does not lose its generality, because it can be shifted to a power series around any point $z = z_0$ by replacing z with $z' = z - z_0$. The following theorem is valid for the convergence and divergence of power series.

Theorem 3.1. *If*

$$f(z) = a_0 + a_1 z + a_2 z^2 + \cdots + a_n z^n + \cdots \quad (3.3)$$

converges at $z = z_0$, the series (3.3) absolutely converges at each point where $|z| < |z_0|$. Also, series (3.3) converges uniformly within a circle centered *at the origin $z = 0$ with a radius of any arbitrary value $\rho < |z_0|$ (uniform convergence on compact sets).*[1]

Proof. If the series converges at $z = z_0$, it follows from Th. 1.6 that

$$\lim_{n\to\infty} a_n z_0{}^n = 0.$$

Thus, for any positive number M, it is possible to choose an appropriate natural number N, and make

$$|a_n z_0{}^n| < M \quad \text{for all } n > N.$$

For such n,

$$|a_n z^n| = |a_n z_0{}^n| \cdot \left|\frac{z}{z_0}\right|^n < M \left|\frac{z}{z_0}\right|^n.$$

$\sum |z/z_0|^n$ is a geometric series and converges for all z of $|z/z_0| < 1$. Therefore, the series (3.3) absolutely converges when $|z/z_0| < 1$.

Next, consider a circle of radius ρ ($\rho < |z_0|$) centered at the origin, in which, for all points z,

$$|a_n z^n| \le |a_n|\rho^n = |a_n z_0{}^n| \left(\frac{\rho}{|z_0|}\right)^n < M \left(\frac{\rho}{|z_0|}\right)^n.$$

The series $\sum M(\rho/|z_0|)^n$ converges (regardless of the value of z), so $\sum |a_n z^n|$ converges uniformly and absolutely. Thus, the series converges uniformly in the circular domain of a radius ρ. $\qquad\square$

Example 3.1. Letting x be a real number, the series

$$1 + x + x^2 + \cdots + x^n + \cdots$$

converges to $1/(1-x)$ at $|x| < 1$. Therefore,

$$1 + z + z^2 + \cdots + z^n + \cdots$$

absolutely converges at $|z| < 1$. This series is equal to $1/(1-z)$.

---

[1]"**Uniform convergence on compact sets** ("compact convergence", or "uniform convergence in the wider sense") in the domain D" means that the series does not converge uniformly in the domain D, but it converges uniformly in any arbitrary bounded closed domain in D.

3.2.2 *Convergence radius*

Definition 3.2 (Convergence circle and convergence radius). According to Th. 3.1, a circle exists on the complex plane in which a series $\sum c_n z^n$ absolutely converges at all points inside it, and diverges at all points outside it. This circle is called the **convergence circle**, and its radius is called the **convergence radius**.

Remark 3.1. It is important to note that Th. 3.1 says nothing about whether the series converges or diverges at points **on** the convergence circle. There are many examples, such as where at some points on the convergence circle the series converges while at others diverges, or where at all points on the circle the series diverges.

The convergence radius can be evaluated by the following method.

Theorem 3.2 (Cauchy–Hadamard theorem). *The convergence radius r of a power series $\sum a_n z^n$ is[2]*

$$\frac{1}{r} = \varlimsup_{n \to \infty} \sqrt[n]{|a_n|}. \tag{3.4}$$

Here, it is necessary to understand that the convergence radius r is 0 when the right-hand side of Eq. (3.4) is infinite (∞), and the convergence radius is infinite when the right-hand side is 0.

Proof. If $|z| < r$, ρ can be selected such that $|z| < \rho < r$. For this ρ, we can see

$$\frac{1}{\rho} > \frac{1}{r} = \varlimsup_{n \to \infty} |a_n|^{1/n}.$$

So if we consider an N that is sufficiently large, from the definition of the superior limit, then for all $n > N$,

$$|a_n|^{1/n} < \frac{1}{\rho}$$

---

[2]In the sequence of real numbers $\{x_n\}$, when there is an infinite number of points arbitrarily close to a point X, that the point X is called the **accumulation point**. To put it another way, the convergence value of the appropriate subsequence $\{x_{n_j}\}$ of the sequence of real numbers $\{x_n\}$ is the accumulation point. The upper limit of the set of accumulation points is represented by $\varlimsup_{n \to \infty} x_n$ or $\limsup_{n \to \infty} x_n$, and is called the **superior limit**. Therefore, the number of points x_n larger than the superior limit is only finite. The lower limit of the set of accumulation points is represented by $\varliminf_{n \to \infty} x_n$ or $\liminf_{n \to \infty} x_n$, and is called the **inferior limit**. From these definitions, the existence of a limit value of $\{x_n\}$ means that the superior limit and inferior limit exist and that they coincide with each other.

is true. Thus,

$$|a_n z^n| < \frac{|z^n|}{\rho^n} \qquad (n > N).$$

As $|z| < \rho$, the series $\sum (|z^n|/\rho)^n$ converges. Therefore, $\sum a_n z^n$ absolutely converges.

Conversely, if $|z| > r$,

$$\frac{1}{|z|} < \frac{1}{r} = \varlimsup_{n \to \infty} |a_n|^{1/n}$$

is the case, so if the subsequence $\{a_{n_i}\}$ is selected appropriately, we can get

$$\frac{1}{|z|} < |a_{n_i}|^{1/n_i}.$$

Thus, there are an infinite number of terms that satisfies

$$|a_{n_i} z^{n_i}| > 1.$$

Therefore, $\sum a_n z^n$ diverges.   □

When determining the convergence radius, it is not always easy to use Cauchy–Hadamard theorem. The following theorem may be used effectively in many cases.

Theorem 3.3. *For the power series $\sum a_n z^n$, if the limit*

$$r = \lim_{n \to \infty} \left| \frac{a_n}{a_{n+1}} \right| \qquad (3.5)$$

exists, r is the convergence radius.

Proof. If the limit

$$r = \lim_{n \to \infty} \left| \frac{a_n}{a_{n+1}} \right|$$

exists, then

$$\varlimsup_{n \to \infty} \left| \frac{a_n}{a_{n+1}} \right| = \varliminf_{n \to \infty} \left| \frac{a_n}{a_{n+1}} \right| = \lim_{n \to \infty} \left| \frac{a_n}{a_{n+1}} \right| = r$$

is true. Therefore, taking each term to be $u_n = a_n z^n$,

$$\lim_{n \to \infty} \left| \frac{u_{n+1}}{u_n} \right| = \lim_{n \to \infty} \left| \frac{a_{n+1} z^{n+1}}{a_n z^n} \right| = |z| \lim_{n \to \infty} \left| \frac{a_{n+1}}{a_n} \right| = \frac{|z|}{r}.$$

Except for the first finite number of terms, the series $\sum |u_n|$ is a geometric series in which the geometric ratio is less than 1 if $|z|/r < 1$, and is greater than 1 if $|z|/r > 1$. From above, it converges because $\sum u_n$ absolutely converges if $|z| < r$. Furthermore, if $|z| > r$, $\lim_{n \to \infty} u_n \neq 0$, so $\sum u_n$ diverges.

□

Example 3.2. (Termwise differentiation of convergent series) When the convergence radius of the series $\sum_{n=1}^{\infty} a_n z^n$ is written as r, the convergence radius of the series $\sum_{n=1}^{\infty} n a_n z^{n-1}$ is r as well.

Proof. If $|z| < r$, one can choose a positive number r_1 with $|z| < r_1 < r$. If series $\sum_{n=1}^{\infty} a_n z^n$ converges, it follows that a number M exists for which $|a_n r_1^n| < M$ (bounded). Thus, $|a_n| < M r_1^{-n}$. Therefore, we get

$$\sum_{n=1}^{\infty} n |a_n| |z|^{n-1} < \frac{M}{|z|} \sum_{n=1}^{\infty} n \frac{|z|^n}{r_1^n}.$$

From Th. 3.3, $\sum n z^n / r_1^n$ converges if $|z|/r_1 < 1$, so above series $\sum n |a_n| |z|^{n-1}$ also converges if $|z|/r_1 < 1$. Therefore, because $r_1 < r$, the series $\sum_{n=1}^{\infty} n |a_n| |z|^{n-1}$ converges if $|z|/r < 1$.

Next, in the case of $|z| > r$, $n |a_n| |z|^{n-1} > |a_n| |z|^n (1/|z|)$. Then, if $\sum_{n=1}^{\infty} a_n z^n$ diverges, so does $\sum_{n=1}^{\infty} n a_n z^{n-1}$ ($\lim_{n \to \infty} (n|a_n|)^{1/n} = \lim_{n \to \infty} |a_n|^{1/n}$). $\qquad \square$

Example 3.3. (Termwise integration of convergent series) If the convergence radius of the series $\sum_{n=1}^{\infty} a_n z^n$ is r, the convergence radius of $\sum_{n=1}^{\infty} a_n z^{n+1}/(n+1)$ is also r.

We can also show this, following the example above (Example 3.2). The integral of a complex function has not yet been defined, but as we will describe later, the integral of z^n is $z^{n+1}/(n+1)$. Therefore, this example tells that a termwise integration can be performed for the integral of a convergent series.

3.3 Exponential functions, trigonometric functions, and hyperbolic functions

3.3.1 *Exponential functions*

Exponential functions have already been described in Sec. 1.2.3. Here we will discuss the differentiation of exponential functions together with related functions, and prepare for the later discussion of logarithmic functions.

Definition 3.3 (Exponential functions). The exponential function $w(z) = \exp z$ is defined by the power series.[3]

$$\exp z = \sum_{n=0}^{\infty} \frac{z^n}{n!}. \tag{3.6}$$

Since $(n + 1)!/n! \to \infty$ $(n \to \infty)$, according to Th. 3.2, the convergence radius of an exponential function is infinite. Therefore, the exponential function is holomorphic at all points on the complex z plane except $z = \infty$.

Differentiation of exponential functions: Differentiating the above series for each term produces

$$\frac{\mathrm{d}}{\mathrm{d}z} \exp z = \exp z. \tag{3.7}$$

This is the differentiation rule of exponential functions.

The exponential law: The exponential law is one important property of exponential functions:

$$\exp(z_1 + z_2) = \exp z_1 \cdot \exp z_2. \tag{3.8}$$

We have already used Definition 3.3 to prove this in Example 1.8.

From the definition of a power series, Euler's formula

$$\exp \mathrm{i}x = \cos x + \mathrm{i} \sin x \tag{3.9}$$

can be derived immediately. The following equation can also be derived. From the definition of a power series, Euler's formula

$$\exp z = \mathrm{e}^x (\cos y + \mathrm{i} \sin y). \tag{3.10}$$

It is also clear that the exponential function is a function with a period of $2\pi\mathrm{i}$ for z.

$$\exp(z + 2m\pi\mathrm{i}) = \exp z.$$

3.3.2   *Trigonometric functions and hyperbolic functions*

Trigonometric functions can simply be generalized as well, according to the generalization of exponential functions. Euler's formula (1.28)

$$\mathrm{e}^{\mathrm{i}\theta} = \cos \theta + \mathrm{i} \sin \theta$$

---

[3] $\exp z$ is normally written as e^z. On the other hand, if we follow the definition of general exponentiation a^z (where a is a complex number), as given in Sec. 3.5.1, e^z differs from the $\exp z$ defined here in the multi-valuedness of the function. The details are described in Sec. 3.5.

and its complex conjugate equation,

$$e^{-i\theta} = \cos\theta - i\sin\theta$$

can be combined to produce

$$\sin x = \frac{e^{ix} - e^{-ix}}{2i}, \qquad \cos x = \frac{e^{ix} + e^{-ix}}{2}.$$

So that equations that hold for a real number x also hold for a complex number z, the exponential function of a complex number is defined as follows.

Definition 3.4 (Trigonometric functions). Sine and cosine functions are defined as

$$\sin z = \frac{e^{iz} - e^{-iz}}{2i} \tag{3.11a}$$

and

$$\cos z = \frac{e^{iz} + e^{-iz}}{2}. \tag{3.11b}$$

Their properties can easily be known from those of exponential functions. For example, it is known that the points where the value becomes 0 (zero) for these functions are only on the real axis, and that the only singular point is the point at infinity.

Euler's formula, addition theorem: In trigonometric functions of complex numbers, Euler's formula, the addition theorem for trigonometric functions, and other functions, are valid as they stand.

$$e^{iz} = \cos z + i\sin z \qquad \text{(Euler's formula)} \tag{3.12a}$$

$$\sin(z_1 + z_2) = \sin z_1 \cos z_2 + \cos z_1 \sin z_2 \quad \text{(Addition theorem 1)} \tag{3.12b}$$

$$\cos(z_1 + z_2) = \cos z_1 \cos z_2 - \sin z_1 \sin z_2 \quad \text{(Addition theorem 2)} \tag{3.12c}$$

$$\sin^2 z + \cos^2 z = 1. \tag{3.12d}$$

The proof is easy, and readers should attempt it for themselves.

Hyperbolic functions, as defined below, are closely related to trigonometric functions.

Definition 3.5 (Hyperbolic functions). The following functions

$$\sinh z = \frac{e^z - e^{-z}}{2} \tag{3.13a}$$

and

$$\cosh z = \frac{e^z + e^{-z}}{2} \tag{3.13b}$$

are called hyperbolic sine, and hyperbolic cosine functions, respectively, and they only have zero points on the imaginary axis. Their only singular point is the point at infinity.

Trigonometric functions and hyperbolic functions can be substituted for each other by replacing z for iz. That fact can be demonstrated immediately from the definitions of trigonometric functions, and hyperbolic functions.

$$\cos z = \cosh(iz), \qquad \cos(iz) = \cosh z, \tag{3.14a}$$

$$i\sin z = \sinh(iz), \qquad \sin(iz) = i\sinh z. \tag{3.14b}$$

Addition theorem for hyperbolic functions, etc.: The addition theorem for hyperbolic functions, and other relationships are as follows:

$$\sinh(z_1 + z_2) = \sinh z_1 \cosh z_2 + \cosh z_1 \sinh z_2, \tag{3.15a}$$

$$\cosh(z_1 + z_2) = \cosh z_1 \cosh z_2 + \sinh z_1 \sinh z_2, \tag{3.15b}$$

$$\cosh^2 z - \sinh^2 z = 1. \tag{3.15c}$$

These can be immediately demonstrated from the definitions, so they should be attempted by readers.

3.4   Logarithmic Functions

If the variable is a real number, the variable x in the logarithmic function $\log x$ must be positive. That is the case because the logarithmic function $y = \log x$ is defined as an inverse function of the power e^y where 'e' is the base of the natural logarithm, Napier number e $= 2.71828\cdots$. We will now consider a logarithmic function that takes a complex number as its variable. The logarithmic function is defined as the inverse function of the exponential function. Once one understands the multi-valuedness of the logarithmic function, that gives a complete understanding of one of the key concepts of complex function theory.

3.4.1   *Definition and principal values of logarithmic functions*

Definition 3.6 (Logarithmic functions). The logarithmic function is defined as the inverse function of the exponential function exp.

$$w = \log z \iff z = \exp w. \tag{3.16}$$

Using the polar representation $z = re^{i\theta} = e^{\ln r}e^{i\theta}$, and paying attention to $e^{i\theta} = e^{i(\theta + 2n\pi)}$ (where n is 0 or an integer),[4] we obtain

$$z = e^{\ln r}e^{i(\theta + 2n\pi)} = e^w = e^{\log z}.$$

---

[4]In this book, we distinguish the logarithmic function log, which takes complex numbers as variables, from the logarithmic function whose variable should be a positive number and whose value is defined definitely to be a real number. The latter is written as 'ln'. $\ln x$ is a single-valued function with the real number x as its variable.

So

$$\log z = \ln r + i(\theta + 2n\pi) = \ln |z| + i(\text{Arg}\, z + 2n\pi) \qquad (3.17)$$

would be a natural result. Equation (3.17) is the definition of a logarithmic function.

In Eq. (3.17), n is 0 or a positive or negative integer, so the function value takes an infinite number of different values according to the value of n. Therefore, the logarithmic function $\log z$ is an infinitely multi-valued function. The expression

$$z = e^{\log z}$$

is often useful.

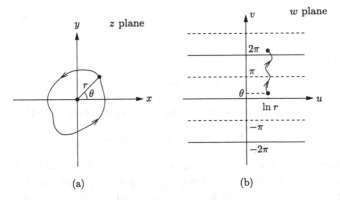

(a)   (b)

Fig. 3.2 Logarithmic functions. (a) When z moves around the origin on the complex z plane, (b) the value of $w = \log z$ on the complex w plane changes by $2\pi i$.

When the variable z moves around the origin, as shown in Fig. 3.2(a), its argument increases (counterclockwise) or decreases (clockwise) by 2π. When that happens, the value of the logarithmic function $\log z$ changes by $2\pi i$. Thus, when the point z moves counterclockwise (clockwise) around the origin on the complex z plane, $\log z$ moves parallel to the imaginary axis on the complex plane, in the positive (negative) direction. When z moves along a closed curve on the z plane, if the origin locates within the domain bounded by that curve, the z argument does not return to its original value, but changes by $+2\pi$ or -2π. Therefore, the curve along which $\log z$ moves is not closed, but only moves by $\pm 2\pi i$. On the other hand, if the closed curve on the z plane does not move around the origin, $\log z$ returns to its original value when z returns to its original position. Therefore, the trajectory of

$\log z$ draws a closed curve. This means that the key to the multi-valuedness of the logarithmic function is the origin on the complex z plane (whether or not the trajectory of z traverses around the origin $z = 0$). That is the case because here $(z = 0)$, the argument of z becomes indeterminate. A point that becomes the source of a multi-valued function, like the origin, in this case, is called a **branch point**. The branch point of a logarithmic function, in particular, is called a **logarithmic branch point**.

To uniquely determine the value of a logarithmic function, it is necessary to define the principal value, as stated below.

Definition 3.7 (Principal value of logarithmic function). The imaginary part of a logarithmic function, limited to between $-\pi$ and π, is called the principal value of a logarithmic function, and is written "Log".

$$\mathrm{Log}\, z = \ln |z| + \mathrm{i}\theta = \ln |z| + \mathrm{i}\,\mathrm{Arg}\, z \qquad (\theta = \mathrm{Arg}\, z \; : \; -\pi < \theta \leq \pi).$$
$$(3.18)$$

This means

$$\log z = \mathrm{Log}\, z + 2n\pi\mathrm{i}.$$

The following addition theorem is valid for a logarithmic function.

$$\log z_1 z_2 = \log z_1 + \log z_2.$$
$$(3.19)$$

In this case (3.19), as stated in Sec. 1.2.2, the infinite number of complex numbers indicated by both sides are equal as a set. To put it another way, both sides of Eq. (3.19) match as multi-valued functions. Looking at the function value, the difference is $2n\pi$ $(n = 0, \pm 1, \pm 2, \cdots)$.

Differential formula for logarithmic function:

$$\frac{\mathrm{d}}{\mathrm{d}z} \log z = \frac{1}{z}.$$
$$(3.20)$$

The proof of this differential formula is easy. Differentiating both sides of $z = \exp(\log z)$ by z produces

$$1 = \frac{\mathrm{d}\log z}{\mathrm{d}z} \frac{\mathrm{d}e^{\log z}}{\mathrm{d}\log z} = \frac{\mathrm{d}\log z}{\mathrm{d}z} e^{\log z} = \frac{\mathrm{d}\log z}{\mathrm{d}z} z.$$

3.4.2 *Multi-valuedness of logarithmic function and Riemann surface*

A logarithmic function is an infinitely multi-valued function. Is there any way to make one value of z correspond to one value of the function $\log z$?

For one value of z on the complex plane, there is an infinite number of corresponding values of $w = \log z$, and it is not possible to uniquely determine the relationship of that correspondence. When z is determined initially, and the value of $\log z$ is set at the same time, subsequent continuous variation of z corresponds to continuous variation in the value of $w = \log z$. Therefore, the value of $w = \log z$ is uniquely determined. However, in that case, it is always necessary to record how many times z has moved around the branch point ($z = 0$).

To avoid this kind of complexity, let us prepare several z planes, and make $\log z$ values correspond to one z plane each. Thus, if z does not move around the origin, we can stay in the same z plane. Let us name this z plane 0th z plane. Then, if z moves n times counterclockwise around the origin, we can visit nth z plane ($2n\pi \leq \arg z < 2(n+1)\pi$); if it rotated n times clockwise around the origin we visit $(-n)$th z plane ($-2n\pi \leq \arg z < -2(n-1)\pi$). In that way, we can consider an infinite number of z planes, from $n = -\infty$ to $n = +\infty$. Connecting these planes definitely, we will move naturally from one z plane to the next, every time z moves around the origin. In that way, when it is determined which z plane z is in, the value of the function $\log z$ is also determined. The z planes, prepared in this way to eliminate the multi-valuedness of a function, are called **Riemann surface**.

If the variable transformation $z = 1/\zeta$ is applied to the logarithmic function $\log z$, it clearly leads to (logarithmic) branch point at $\zeta = 0$ ($z = \infty$). On the z plane, cutting the part of the real axis between the two branch points $z = 0$ and $z = \infty$ (such as the right half of the real axis),

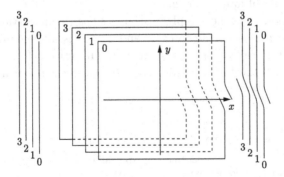

Fig. 3.3 An infinite number of sheets of Riemann surface for a logarithmic function. The left and right sides of the diagram show the appearance of the Riemann surface when viewed from those sides.

and the part of the $\arg z = 2(n+1)\pi$ on the nth z plane and that of the $\arg z = 2(n+1)\pi$ on the $(n+1)$th z plane are connected (Fig. 3.3). Once that is done with an infinite number of z planes, we can make one point on a z plane be in one-to-one correspondence with one point on the $w = \log z$ plane, and the infinite number of z planes are continuously connected.

3.5 General exponential functions and multi-valuedness

We have so far considered integer powers of the complex number z, and rational number powers (Chapter 1). We will now define not only arbitrary real number powers, but also complex number powers of complex numbers, *i.e.* the general exponential function.

3.5.1 *Definition of general exponential function*

Definition 3.8 (General exponential function). The function a^b, in which a and b are complex numbers, is defined as[5]

$$a^b = \exp(b \log a) = \exp[b(\ln|a| + \mathrm{i}\arg a)]. \tag{3.21}$$

Since $\arg a = \operatorname{Arg} a + 2n\pi$, so from the above definition,

$$a^b = \exp[b(\ln|a| + \mathrm{i}\operatorname{Arg} a + \mathrm{i}2n\pi)]. \tag{3.22}$$

Handling $\exp b(\mathrm{i}2n\pi)$ with care, a^b is a single-valued function, provided b is an integer. If b is a rational number p/q, a^b is generally a q-valued function. If b is otherwise an irrational or complex number, a^b is an infinitely multi-valued function. The multi-valuedness of a^b is caused by the multi-valuedness of $\log a$ or $\arg a$.

The function z^a of the complex variable z has the multi-valuedness described above if a is not an integer, and $z = 0$ is a singularity (branch point). $z = 0$ is called the **algebraic branch point** when a is a rational number, and the **logarithmic branch point** when a is an irrational

---

[5]Following this definition produces

$$\mathrm{e}^z \equiv \exp(z \log \mathrm{e}) = \exp[z(\ln \mathrm{e} + \mathrm{i}2n\pi)]$$
$$= \exp z(1 + \mathrm{i}2n\pi) = \exp z \cdot \exp(\mathrm{i}2n\pi z).$$

Therefore, e^z and the exponential function $\exp z$ differ. However, in general, we take only the one with $n = 0$ for e^z above, and let it be

$$\mathrm{e}^z \equiv \exp z.$$

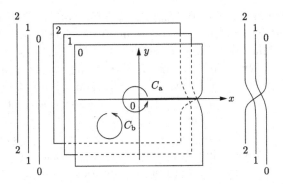

Fig. 3.4   Riemann surface for the power function $w = z^{1/3}$. Refer to the text for C_a and C_b.

number or is not a real number. On the other hand, "once the value of (the imaginary part of) $\log a$ is determined", a^z has a uniquely determined value and is, therefore, a single-valued function.

Example 3.4. Once the value of $\log z$ is uniquely determined, the relation

$$z^a z^b = z^{a+b}, \qquad (z^a)^b = z^{ab}$$

is valid.

The differentiation of general exponential function is obtained as follows, employing the differentiation of a composite function and an exponential function.

Differential formula for power functions:

$$\frac{d}{dz}z^a = \frac{d}{dz}e^{a \log z} = \frac{da \log z}{dz}e^{a \log z} = \frac{a}{z}z^a = az^{a-1}, \qquad (3.23a)$$

$$\frac{d}{dz}a^z = \frac{d}{dz}e^{z \log a} = \log a\, e^{z \log a} = a^z \log a. \qquad (3.23b)$$

3.5.2   *Mapping of multi-valued function $w = z^{1/n}$, and Riemann surface*

Let us consider a function $w = z^{1/n}$. This is an n-valued function, and $z = 0$ is the singular point. Thus, a rotation through 2π around $z = 0$ does not return w to its original value (circular path C_a in Fig. 3.4). However, rotating by 2π around other points ($z \neq 0$) at a finite distance from the origin (circular path C_b in Fig. 3.4) restores w to its original value. $z = 0$ is an algebraic **branch point**.

$w = z^{1/n}$ is an n-valued function, so after rotating n times around $z = 0$, the function w returns to its original value. Let us prepare n sheets of complex z planes and consider z with different arguments on each plane. $0 \le \arg z < 2\pi$, $2\pi \le \arg z < 4\pi$, and $4\pi \le \arg z < 6\pi$ are mapped to $0 \le \arg w < 2\pi/n$, $2\pi/n \le \arg w < 4\pi/n$, and $4\pi/n \le \arg w < 6\pi/n$, respectively. Thus, it is necessary to set that visiting n sheets of z planes by rotating around the origin n times brings w back to the original point on the w plane as well. Note that other than $z = 0$, $z = \infty$ is also a branch point of $w = z^{1/n}$. Certainly, if we set $z = 1/\zeta$, we get $w = \zeta^{-1/n}$, so $\zeta = 0$, which is to say, $z = \infty$, is also a branch point of $w = z^{1/n}$.

For simplicity, let us consider $w = z^{1/3}$. Choose any arbitrary curve connecting the two branch points $z = 0$ and $z = \infty$, so cut across the complex z plane. Now, let us choose a straight line connecting $x = 0$ and $x = \infty$ (the part of $x \ge 0$) on the real axis. Cutting here, connect the part $\arg z = 2\pi - 0$ (the part below the cut on the 0th z plane), and the part $\arg z = 2\pi + 0$ (the part above the cut on the 1st z plane), the part $\arg z = 4\pi - 0$ (the part below the cut on the 1st z plane) and the part $\arg z = 4\pi + 0$ (the part above the cut on the 2nd z plane), as shown in Fig. 3.4. Furthermore, the part $\arg z = 6\pi - 0$ (the part below the cut on the 2nd z plane) and the part $\arg z = 0+$ (the part above the cut on the 0-th z plane) are also connected. The final connection passes through the other two planes, but we only have to imagine it conceptually, so there is no need to worry. As Fig. 3.4 shows, the connection of three z planes into one is complete, and z and w can be made to correspond on a one-to-one basis.

3.6   Infinite product

3.6.1   *Definition of infinite product, convergence, and divergence*

Definition 3.9. Consider creating the product of infinite terms $\prod\limits_{n=1}^{\infty}(1+u_n)$ from an infinite sequence $\{u_n\}_{n=1}^{\infty}$.

$$(1 + u_1)(1 + u_2)(1 + u_3)\cdots(1 + u_n)\cdots = \prod_{n=1}^{\infty}(1 + u_n) \qquad (3.24)$$

is called an **infinite product**. Here, we assume that no value $(1 + u_n)$ becomes 0, and that the limit value of Eq. (3.24) is not 0.

Definition 3.10 (Convergence and divergence of infinite product).
When the infinite product (3.24) is a non-zero unique finite value, it is said
to converge. When it does not converge (which is to say, it does not take
any unique finite value, or the value becomes ∞ or 0), it is said to diverge.

Definition 3.11 (Absolute convergence of infinite product). When
the infinite series $\sum_{n=1}^{\infty} \text{Log}\,(1+u_n)$ absolutely converges, the infinite product
(3.24) is said to absolutely converge.

Theorem 3.4. *The necessary and sufficient condition for the infinite prod-*
uct $\prod_{n=1}^{\infty}(1+u_n)$ to converge is that the infinite series $\sum_{n=1}^{\infty} \text{Log}\,(1+u_n)$ con-
verges.

Proof. Consider the partial product p_N;

$$p_N = \prod_{n=1}^{N}(1+u_n) = \exp\left[\sum_{n=1}^{N} \text{Log}\,(1+u_n)\right]. \tag{3.25}$$

Here, the logarithmic function takes a principal value. As for finding the
limit value of the exponential function for an arbitrary series $\{u_n\}_{n=1}^{\infty}$,
if $\lim_{n\to\infty} u_n$ exists, the continuity of the exponential function means that
$\exp\left(\lim_{n\to\infty} u_n\right) = \lim_{n\to\infty} \exp u_n$. So, if we take $N \to \infty$ in Eq. (3.25), we
obtain

$$\lim_{N\to\infty} p_N = \prod_{n=1}^{\infty}(1+u_n) = \exp\left[\sum_{n=1}^{\infty} \text{Log}\,(1+u_n)\right]. \tag{3.26}$$

Therefore, if the term $\sum_{n=1}^{\infty} \text{Log}\,(1+u_n)$ in Eq. (3.26) converges, the left-hand
side exists. Since the logarithmic function has a principal value here, this
condition can be satisfied for $|u_n| \to 0$.

Conversely, let's assume the left-hand side of Eq. (3.26) to converge to
p. Taking the logarithm of both sides of Eq. (3.26) produces

$$\text{Log} \prod_{n=1}^{\infty}(1+u_n) = \sum_{n=1}^{\infty} \text{Log}\,(1+u_n), \tag{3.27}$$

so the right-hand side converges. □

Theorem 3.5. *The necessary and sufficient condition for the infinite prod-*
uct $\prod_{n=1}^{\infty}(1+u_n)$ to absolutely converge is that the series $\sum_{n=1}^{\infty} u_n$ must abso-
lutely converge. Here, we assume $u_n \neq -1$.

Proof. To prove this theorem, it is enough to say that the necessary and sufficient condition for the convergence of $\sum\limits_{n=1}^{\infty} |\text{Log}(1 + u_n)|$ is the convergence of the series $\sum\limits_{n=1}^{\infty} |u_n|$. When $u_n \to 0$, it produces $[\text{Log}\,(1 + u_n)]/u_n \to 1$. Then, for any arbitrary $\varepsilon > 0$, $(1 - \varepsilon) < [\text{Log}\,(1 + u_n)]/|u_n| < (1 + \varepsilon)$, provided that an enough large n is used. Therefore, it follows that if $u_n \to 0$, $|u_n|(1 - \varepsilon) < |\text{Log}\,(1 + u_n)| < |u_n|(1 + \varepsilon)$ can be taken for these ε and n. This means simultaneous absolute convergence of $\prod\limits_{n=1}^{\infty} (1 + u_n)$ and $\sum\limits_{n=1}^{\infty} u_n$. In other words, the necessary and sufficient condition for one of $\prod(1 + u_n)$ and $\sum u_n$ to absolutely converge is that for the other to absolutely converge. $\qquad\square$

3.6.2   *Example of infinite product (infinite product representation of* sin z, cos z*)*

Infinite products appear in various places. It is appropriate to deal with them when we understand the analytical properties of complex functions a little further, and it would be difficult to explain them fully at this stage. Here, we will only explain the key points and several results, which are sufficient to advance an overall understanding. (See for details Sec. 10.1.2.)

Consider a trigonometric function $\sin z$ that takes a complex number z as its variable. The exponential function $\exp z$ can be written as a power series (Example 1.8). Also, as $\sin z = (e^{iz} - e^{-iz})/(2i)$, $\sin z$ can be expressed as an infinite series of a polynomial. In fact, this is the Taylor expansion of $\sin z$,

$$\sin z = \sum_{n=0}^{\infty} (-1)^n \frac{z^{2n+1}}{(2n + 1)!}. \tag{3.28}$$

Furthermore, the exponential function never diverges anywhere other than the point at infinity (holomorphy), and $n\pi$ is the only zero point. Then, $\sin z$ has $(z \pm n\pi)$ or $[1 \pm (z/n\pi)]$ as factors. Because $z = 0$ is also a zero point, which means that z also exists as a factor. This suggests that we would obtain[6]

$$\sin z = Cz \prod_{n=1}^{\infty} \left(1 - \frac{z^2}{n^2 \pi^2}\right).$$

---

[6]There is another possibility to have a factor of $\exp(h(z))$, where $h(z)$ is a holomorphic function. For detail, see Sec. 10.1.2.

C is a constant to be determined below. Comparison between the coefficient of z when the right-hand side is expanded and the coefficient of z for the first term of the Taylor expansion reveals that $C = 1$. This produces the following infinite product representation of $\sin z$,

$$\frac{\sin z}{z} = \prod_{n=1}^{\infty} \left(1 - \frac{z^2}{n^2 \pi^2} \right). \qquad (3.29)$$

The following equation can be obtained similarly.

$$\cos z = \prod_{n=1}^{\infty} \left[1 - \frac{z^2}{(n - \frac{1}{2})^2 \pi^2} \right]. \qquad (3.30)$$

Chapter 4

Conformal Transformation

A variety of beautiful rules exist in mapping by holomorphic functions. For example, a very small domain in the original space can be mapped by a holomorphic function to a similar very small domain in the image space. This property is called "conformal". Also, the real part and the imaginary part of a holomorphic function are harmonic functions (solutions of Laplace equation), so mapping by a holomorphic function (called conformal transformation, conformal mapping) is applied in a variety of fields in physics and engineering.

4.1   Definition of conformal transformation

Let us consider, for the holomorphic function $w = f(z)$ in the domain D on the complex z plane, three adjacent points z_i $(i = 0, 1, 2)$ in D (Fig. 4.1(a)). The smooth curve joining z_0 and z_i $(i = 1, 2)$ is mapped, because of the continuity (in the holomorphic domain D) of the function f, to the smooth curve joining w_0 and w_i $(i = 1, 2)$ in the same way on the complex w plane.

In that case, by using the polar coordinates on the z plane and the w plane, near the corresponding centers z_0, w_0, one can obtain

$$z_1 - z_0 = r_1 e^{i\theta_1}, \qquad z_2 - z_0 = r_2 e^{i\theta_2},$$
$$w_1 - w_0 = \rho_1 e^{i\phi_1}, \qquad w_2 - w_0 = \rho_2 e^{i\phi_2}. \tag{4.1}$$

From the holomorphy of $w = f(z)$

$$f'(z_0) = \lim_{z_1 \to z_0} \frac{w_1 - w_0}{z_1 - z_0} = \lim_{z_2 \to z_0} \frac{w_2 - w_0}{z_2 - z_0}, \tag{4.2}$$

and this is rewritten in polar coordinate as

$$f'(z_0) = \lim_{z_1 \to z_0} \frac{\rho_1}{r_1} e^{i(\phi_1 - \theta_1)} = \lim_{z_2 \to z_0} \frac{\rho_2}{r_2} e^{i(\phi_2 - \theta_2)},$$

49

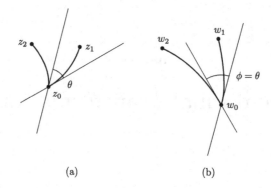

(a)   (b)

Fig. 4.1   Three points z_i ($i = 0, 1, 2$) are mapped to three points w_i ($i = 0, 1, 2$) by $w = f(z)$, and the conformal transformation assures that $\phi = \theta$.

or

$$\frac{\rho_1}{r_1}e^{i(\phi_1 - \theta_1)} = f'(z_0) + \gamma_1, \qquad \frac{\rho_2}{r_2}e^{i(\phi_2 - \theta_2)} = f'(z_0) + \gamma_2. \qquad (4.3)$$

Here, γ_1 and γ_2 are complex numbers that approach 0 more quickly than r_1, r_2, ρ_1, ρ_2, etc. when $z \to z_0$. Therefore, if we let $z_1 \to z_0$ and $z_2 \to z_0$ in Eq. (4.3), we can see

$$\lim_{z_1, z_2 \to z_0} \frac{\rho_2}{\rho_1} = \lim_{z_1, z_2 \to z_0} \frac{r_2}{r_1}, \qquad \lim_{z_1, z_2 \to z_0} (\phi_2 - \phi_1) = \lim_{z_1, z_2 \to z_0} (\theta_2 - \theta_1).$$

The above statement can be expressed in geometric terms as "a very small triangle $\triangle z_0 z_1 z_2$ in the neighborhood of z_0 is mapped to a similarly very small triangle $\triangle w_0 w_1 w_2$ in the neighborhood of w_0". Thus,

$$\triangle z_0 z_1 z_2 \sim \triangle w_0 w_1 w_2. \qquad (4.4)$$

It can be seen immediately from Eq. (4.3) that $|f'(z_0)|$ is the magnification ratio of the similar figure, and $\arg f'(z_0)$ is the rotation angle of the figure $(\phi_1 - \theta_1) = (\phi_2 - \theta_2)$. It is said that "function $f(z)$ is conformal", and that function $f(z)$ is a **conformal transformation** or **conformal mapping**. It can be understood from the above explanation that when a point z moves along a certain path, the domain on the right (left) is mapped to the same right (left) side of the path corresponding to the image curve.

Definition 4.1 (Conformal). When there is a mapping $w = f(z)$ that is defined in the domain D, and the angle formed by any two arbitrary smooth curves passing through the point z_0 is equal, including the sign, to the angle formed by the two image curves passing through the image point w_0, $w = f(z)$ is said to be **conformal** at the point z_0.

Theorem 4.1. *If the function $f(z)$ is holomorphic at z_0 and $f'(z_0) \neq 0$, the mapping $w = f(z)$ is conformal at z_0.*

This is another way of stating the property of conformal transformation. Note, however, that when $f'(z_0) = 0$, the argument of the mapping $w = f(z)$ becomes indefinite at $z = z_0$, so the transformation is not conformal at z_0.

4.2 Simple examples of conformal transformations

Let us examine simple examples of a conformal transformation.

Example 4.1. In the equation

$$w = \frac{1}{z},$$

when $z = re^{i\theta}$, $w = (1/r)e^{-i\theta}$, so the circle centered at $z = 0$ is mapped to a circle centered at $w = \infty$ (which is also the circle centered at $w = 0$), and the group of straight lines extending radially from $z = 0$ is mapped to a group of straight lines concentrating on $w = 0$ (the group of straight lines coming from $z = \infty$). If the point z moves counterclockwise on the circle centered at $z = 0$ on the z plane, the point w moves clockwise on the circle centered at $w = 0$ on the w plane. Therefore, the domain interior (exterior) to the circle on the z plane is mapped to the domain exterior (interior) to the circle on the w plane. A group of orthogonal curves on the z plane is mapped to a group of orthogonal curves on the w plane, and the transformation is conformal, except $z = 0$ (Fig. 4.2).

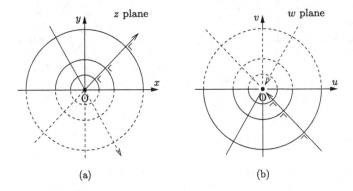

(a)   (b)

Fig. 4.2   Complex z plane and complex w plane produced by the mapping $w = 1/z$. A group of orthogonal curves on the z plane is mapped to a group of orthogonal curves on the w plane, and the mapping is conformal except $z = 0$.

Example 4.2.

$$w = z^2.$$

This mapping can be written, with the real part and the imaginary part separated, as

$$u = x^2 - y^2, \qquad v = 2xy, \qquad (4.5)$$

so the half line $x > 0$, $y = 0$ along the real axis on the z plane is mapped to $u = x^2 > 0$, $v = 0$, and the half line $x = 0$, $y > 0$ along the imaginary axis on the z plane is mapped to $u = -y^2 < 0$, $v = 0$. Also, groups of straight lines parallel to the real axis and imaginary axis on the z plane are each mapped to groups of palabora, as below:

$$x = a \Rightarrow u = a^2 - \left(\frac{v}{2a}\right)^2, \qquad (4.6a)$$

$$y = b \Rightarrow u = \left(\frac{v}{2b}\right)^2 - b^2. \qquad (4.6b)$$

These two groups of palabora intersect perpendicularly everywhere on the w plane (Fig. 4.3). Furthermore, it can be seen that the arcs in the first quadrant of the z plane are mapped to the arcs in the upper half of the w plane. From the above, this function is conformal everywhere except $z = 0$, which is a point of $f'(0) = 0$.

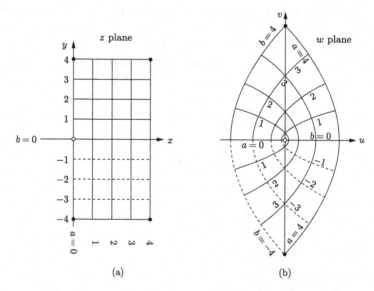

(a)   (b)

Fig. 4.3 $w = z^2$. The group of straight lines parallel to the real axis and imaginary axis on the z plane is mapped to the group of palabora on the w plane, and they intersect perpendicularly everywhere on the w plane.

4.3   Linear transformations

The mapping by linear fractional function shown below is generally important and is also useful as a conformal transformation.

4.3.1   *Entire functions and rational functions*

Where n is 0 or a positive integer, and $a_0, a_1, a_2, \cdots$ are complex constants $(a_0 \neq 0)$

$$P(z) = a_0 z^n + a_1 z^{n-1} + \cdots + a_{n-1} z + a_n \tag{4.7}$$

is called an (nth order) **polynomial**, or a **rational integral function**. The polynomial is holomorphic on the whole complex z plane (all points within a finite distance from the origin). In general, a function that is holomorphic for all points on the complex plane within a finite distance from the origin is called an **entire function** or an **integral function**.

When $P(z)$ and $Q(z)$ are polynomials of z $(Q(z) \not\equiv 0)$,

$$\frac{P(z)}{Q(z)} = \frac{a_0 z^n + a_1 z^{n-1} + \cdots + a_{n-1} z + a_n}{b_0 z^m + b_1 z^{m-1} + \cdots + b_{m-1} z + b_m} \tag{4.8}$$

is called a **rational function**. In particular, a first-order rational function taking a, b, c, d as complex constants,

$$w = \frac{az + b}{cz + d} \qquad (ad - bc \neq 0) \tag{4.9}$$

is called a **linear fractional function**, or just a **linear function**.

4.3.2   *Linear fractional functions and linear transformations*

A mapping by a linear fractional function is called a **linear transformation** (linear mapping), or **Möbius transformation**. Rewriting Eq. (4.9),

$$w = \frac{(bc - ad)/c}{cz + d} + \frac{a}{c}, \tag{4.10}$$

so a linear transformation is a combination of the following three transformations.

(1) $w = z + \alpha$
(2) $w = \beta z$
(3) $w = 1/z$

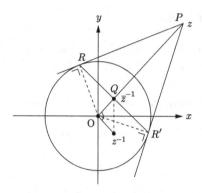

Fig. 4.4 The position $1/z$ and the mirror images $1/\bar{z}$ of z. Tangents drawn to the unit circle from P contact it at R and R'. Lines segments $\overline{RR'}$ and $\overline{OP}$ intersect at Q.

These three transformations are the fundamentals of mapping and are also important as conformal transformations. Transformation (1) is translation on the complex plane, and transformation (2) is enlargement (reduction) and rotation on the complex plane. Transformation (3) requires a little preparation.

In Fig. 4.4, right-angled triangle $\triangle ORQ$ and right-angled triangle $\triangle OPR$ are **similar**, so

$$\frac{\overline{OQ}}{\overline{OR}} = \frac{\overline{OR}}{\overline{OP}}.$$

Therefore, if the circle radius $\overline{OR} = 1$, it follows that $\overline{OQ} \cdot \overline{OP} = 1$. This means that if the complex number z is the point P, the point Q is $\bar{z}^{-1}$. (Points O, P, and Q lie on a straight line, so the arguments of the points P and Q (as complex numbers) are equal. Therefore, the point Q is not z^{-1}, but $\bar{z}^{-1}$.) It is easy to see that if $|z| < 1$, the locations of points P and Q are interchanged in Fig. 4.4. Points P and Q are said to be in the position of a mirror image with respect to the unit circle. $1/z$ can be found from Fig. 4.4, corresponding to transformation (3), in which z (P) is moved to the position of the mirror image (Q), and then to a point symmetrical with respect to the real axis.

The linear fractional function (4.9) is holomorphic and conformal on the entire complex z plane, with an exception of $z = -d/c$. Also, this gives a continuous one-to-one correspondence between points on the z plane and points on the w plane, with exceptions of $z = -d/c$ and $w = a/c$. Points that are fixed through the transformation[1] are roots of $cz^2 + (d-a)z - b = 0$.

---

[1] The point which is mapped to itself $w(z) = z$. It is called a fixed point.

There are generally two such points.

The equations of a straight line or a circle on the z plane (where a and c are real numbers) can be written as

$$a z \bar{z} + \bar{z}_0 z + z_0 \bar{z} + c = 0. \tag{4.11}$$

The linear transformation $w = 1/z$ changes Eq. (4.11) into

$$c w \bar{w} + z_0 w + \bar{z}_0 \bar{w} + a = 0. \tag{4.12}$$

Equation (4.12) is a straight line or a circle on the w plane as well. A general linear transformation consists of, adding to $w = 1/z$, translation, enlargement, reduction, and rotation. That means that a linear transformation maps a straight line or a circle on a plane to a straight line or a circle. This is called **circle-to-circle correspondence**.

4.3.3 *Example of linear transformation (conformal transformation)*

We have already stated that mapping by linear fractional function is a conformal transformation. Let us examine some examples.

Example 4.3. Consider a linear transformation (Im $z_0 > 0$)

$$w = e^{i\alpha} \frac{z - z_0}{z - \bar{z}_0}. \tag{4.13}$$

If we let $z = x$, then $|w| = 1$. Therefore, the real axis of the z plane is mapped to the unit circle on the w plane. Furthermore, the continuity of linear transformation explains that the upper half of the z plane (Im $z > 0$) is mapped to the outside or the inside of the unit circle on the w plane. The point on the z plane corresponding to $w = 0$ is $z = z_0$, so the upper (lower) half of the z plane is mapped to the inside (outside) of the unit circle on the w plane (Fig. 4.5(a) shows the case of Im $z_0 > 0$).

Example 4.4. Consider a linear transformation ($|z_0| < 1$)

$$w = e^{i\alpha} \frac{z - z_0}{\bar{z}_0 z - 1}. \tag{4.14}$$

If $z = \exp(i\theta)$, then $|w| = 1$, so the unit circle of the z plane is mapped to the unit circle on the w plane. Furthermore, the point on the z plane corresponding to $w = 0$ is $z = z_0$, so the inside (outside) of the unit circle of the z plane is mapped to the inside (outside) of the unit circle on the w plane (Fig. 4.5(b) shows the case of $|z_0| < 1$).

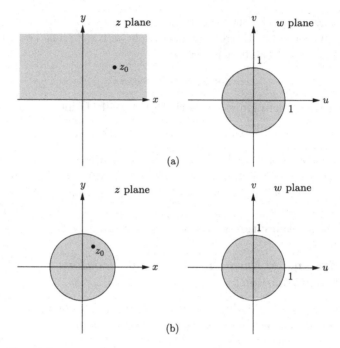

Fig. 4.5 Domains on the z plane, and their corresponding domains on the w plane. (a) Linear transformation $w = e^{i\alpha}(z - z_0)/(z - \bar{z}_0)$ ($\mathrm{Im}\, z_0 > 0$). (b) Linear transformation $w = e^{i\alpha}(z - z_0)/(\bar{z}_0 z - 1)$ ($|z_0| < 1$).

4.4   Harmonic functions and conformal transformations

It was explained in Sec. 2.2.3.1 that a function satisfying Laplace equation is a **harmonic function**. So, both the real part u and the imaginary part v of a holomorphic function are harmonic functions (in two dimensions). Laplace equation appears in various situations, such as the equation for electrostatic potential in a domain with no electrical charge in electromagnetism, and in the equation for the velocity potential in fluid dynamics. Therefore, harmonic functions are very important in the fields of physics and engineering.

4.4.1   *Transformation of Laplace equation by conformal mapping*

As we saw in Eq. (2.18a), the real part and the imaginary part of a holomorphic function are harmonic functions.

If we use the conformal mapping $w = u(x, y) + iv(x, y)$ to perform a variable transformation

$$u = u(x, y), \qquad v = v(x, y) \tag{4.15}$$

and Cauchy–Riemann relation, Laplace operator in (x, y) space is transformed into

$$\frac{\partial^2}{\partial x^2} + \frac{\partial^2}{\partial y^2} = \left[\left(\frac{\partial u}{\partial x} \right)^2 + \left(\frac{\partial u}{\partial y} \right)^2 \right] \left(\frac{\partial^2}{\partial u^2} + \frac{\partial^2}{\partial v^2} \right). \tag{4.16}$$

The proportionality coefficient of the transformation is the enlargement ratio of the conformal mapping:

$$\left(\frac{\partial u}{\partial x} \right)^2 + \left(\frac{\partial u}{\partial y} \right)^2 = |w'(z)|^2. \tag{4.17}$$

From Eq. (4.16), Laplace equation in (x, y) space is transformed into Laplace equation in (u, v) space.

$$\left(\frac{\partial^2}{\partial x^2} + \frac{\partial^2}{\partial y^2} \right) f(x, y) = 0 \Leftrightarrow \left(\frac{\partial^2}{\partial u^2} + \frac{\partial^2}{\partial v^2} \right) f(x(u, v), y(u, v)) = 0. \tag{4.18}$$

Consider a two-dimensional complex function $f = \Phi + i\Psi$ that satisfies Laplace equation. Here, we use the holomorphic function $z = z(w)$ $(w = w(z))$ to map the z plane onto the w plane. The complex function $f(z(w)) = \Phi + i\Psi$ is differentiable for w as well as z, so the function is a holomorphic function for w as well, and from the above discussion, Φ and Ψ are each harmonic functions of u and v ($w = u + iv$). This is the meaning of Eq. (4.18).

4.4.2 *Harmonic functions in electromagnetism and fluid dynamics*

4.4.2.1 *Electromagnetism, electrostatic potential, and lines of electric force*

The **electrostatic potential** $\phi(\mathbf{r})$ in a three-dimensional space with no charge satisfies Laplace equation

$$\left(\frac{\partial^2}{\partial x^2} + \frac{\partial^2}{\partial y^2} + \frac{\partial^2}{\partial z^2} \right) \phi(\mathbf{r}) = 0. \tag{4.19}$$

When a particle with charge q is placed at point $\mathbf{r}$, the force $\mathbf{F}$ acting on it can be expressed by the gradient $\text{grad}\, \phi$ of electrostatic potential.

$$\mathbf{F}(\mathbf{r}) = -q\, \text{grad}\ \phi(\mathbf{r}) = -q \left(\frac{\partial}{\partial x}, \frac{\partial}{\partial y}, \frac{\partial}{\partial z} \right) \phi(\mathbf{r}) \tag{4.20a}$$

The potential gradient with the sign reversed at each point, which is the vector having a direction and size of potential gradient, is called the electric field vector **E**.

$$\mathbf{E}(\mathbf{r}) = -\mathrm{grad}\,\phi(\mathbf{r}) = -\left(\frac{\partial}{\partial x}, \frac{\partial}{\partial y}, \frac{\partial}{\partial z}\right)\phi(\mathbf{r}). \qquad (4.20b)$$

When the arrows representing the electric field vectors are drawn in space and joined together continuously, they are called **lines of electric force**. Lines of electric force follow the direction of the electric field and intersect perpendicularly with the curve $\phi(\mathbf{r}) = $ constant. The density of the lines of electric force is proportional to the strength of the electric field (the size of the electric field vector). Therefore, lines of electric force represent the direction of force acting on a charged particle.

If $\phi(\mathbf{r})$ does not depend on z, any dependence on z in the above discussion can be ignored, and it is sufficient to consider two-dimensional coordinates. The two-dimensional electrostatic potential $\phi(\mathbf{r})$ and electric field vector $\mathbf{E}(\mathbf{r})$ become

$$\left(\frac{\partial^2}{\partial x^2} + \frac{\partial^2}{\partial y^2}\right)\phi(\mathbf{r}) = 0, \qquad (4.21a)$$

and

$$\mathbf{E}(\mathbf{r}) = -\mathrm{grad}\,\phi(\mathbf{r}) = -\left(\frac{\partial}{\partial x}, \frac{\partial}{\partial y}\right)\phi(\mathbf{r}), \qquad (4.21b)$$

respectively. Electrostatic potential $\phi(x, y)$ becomes a (two-dimensional) harmonic function. Also, the gradient of electrostatic potential (the vector with its sign reversed), expresses the electric field vector. If we write the conjugate harmonic function of $\phi(x, y)$ as $\psi(x, y)$, they must satisfy Cauchy–Riemann relation as

$$\frac{\partial}{\partial x}\phi(x, y) = \frac{\partial}{\partial y}\psi(x, y), \qquad \frac{\partial}{\partial y}\phi(x, y) = -\frac{\partial}{\partial x}\psi(x, y), \qquad (4.22)$$

so the gradient of ψ is orthogonal to the gradient of ϕ (*i.e.* grad$\psi \perp$ gradϕ). Therefore, the curve $\psi = $ constant determines the lines of electric force, which intersect perpendicularly with the curve $\phi = $ constant (equipotential curves of electrostatic potential).

4.4.2.2   *Irrotational flow, velocity potential, and stream function*

Letting the vector of the velocity field (the velocity vector) be $\mathbf{v}(\mathbf{r})$, $\omega(\mathbf{r}) = $ rot $\mathbf{v}$ is called **vorticity**, representing the rotation of a vortex flow. When

$\omega = \mathbf{0}$, it is called **irrotational flow**. When $\mathrm{rot}\,\mathbf{v} = \mathbf{0}$, $\mathbf{v}$ can be derived from the function Φ, which is written as $\mathbf{v} = \mathrm{grad}\,\Phi$. This Φ is called the **velocity potential**. Once we write the mass density of the flow as ρ and the time as t, the continuity equation is expressed as

$$\frac{\partial \rho}{\partial t} + \mathrm{div}(\rho \mathbf{v}) = 0. \tag{4.23}$$

In an irrotational and incompressible fluid, ρ is a constant, it produces

$$\mathrm{div}\,\mathbf{v} = 0 \Rightarrow \begin{array}{c} \mathbf{v} = \mathrm{grad}\Phi \\ (\mathrm{div}\,\mathrm{grad}\,\Phi = 0) \end{array} \Rightarrow \left(\frac{\partial^2}{\partial x^2} + \frac{\partial^2}{\partial y^2} + \frac{\partial^2}{\partial z^2} \right) \Phi = 0 \tag{4.24}$$

and the velocity potential Φ is a solution of Laplace equation.

In this situation, consider a two-dimensional flow $\mathbf{v}$ which depends only on x, y, and is independent of z. Velocity potential Φ satisfies the two-dimensional Laplace equation

$$\left(\frac{\partial^2}{\partial x^2} + \frac{\partial^2}{\partial y^2} \right) \Phi = 0 \tag{4.25a}$$

and the velocity field of the flow becomes

$$v_x = \frac{\partial \Phi}{\partial x}, \qquad v_y = \frac{\partial \Phi}{\partial y}. \tag{4.25b}$$

Consider one point A in space, and any arbitrary point P, as shown in Fig. 4.6. Let v_n be the component of $\mathbf{v}$ projected in the direction obtained by rotating the tangent line of the curve C drawn from A to P clockwise

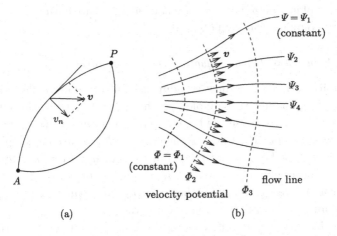

Fig. 4.6 The stream function in an incompressible fluid.

by 90°. In an incompressible fluid, the volume of a fluid passing across the curve C, which joins A and P, is

$$\Psi(P) = \int_A^P v_n \, ds, \qquad (4.26)$$

and that is determined by the point P only, regardless of the curve C. That is the case because, considering any two arbitrary curves C and C', joining A and P, the fact that the flow entering through C cannot be compressed means that it must pass out through C'. This $\Psi(x, y)$ is called the **stream function**. The flow volume passing through the curve $\Psi =$ constant is 0, so $\Psi(x, y) =$ constant provides the flow line.

According to the definition, the differential of the stream function Ψ is the velocity of the flow, $v_n = \partial\Psi/\partial s$, which means that if the stream function is differentiated in a certain direction, the result will be the velocity in the direction turned further 90° clockwise in that position. Therefore, taking the directions of differentiation as the x and y directions, respectively, the velocities in the $-y$, x directions are $-v_y$ and v_x.

$$v_x = \frac{\partial\Psi}{\partial y}, \qquad v_y = -\frac{\partial\Psi}{\partial x}. \qquad (4.27)$$

Comparing the equation using velocity potential for the velocity field of a flow (4.25b) and the equation using the stream function (4.27), it is clear that they are Cauchy–Riemann relation. Thus, the function in which the real part and the imaginary part are Φ and Ψ, respectively,

$$f = \Phi + i\Psi, \qquad (4.28)$$

is a holomorphic function of $z = x + iy$, and Φ and Ψ are harmonic functions. f is called complex velocity potential.

4.4.2.3   *Solution of Laplace equation where boundaries exist*

Consider irrotational flow in a two-dimensional incompressible fluid. We have already seen that the complex potential $f = \Phi + i\Psi$ in this situation has been defined. If we use the holomorphic function $z = z(w)$ $(w = w(z))$ here to map the z plane onto the w plane, Φ and Ψ for the complex potential $f(z(w)) = \Phi + i\Psi$ become harmonic functions of u, v $(w = u + iv)$ in the w plane, respectively. Therefore, in the $w = u + iv$ plane, the velocity field exists for an irrotational flow in which $\Phi(u, v)$, and $\Psi(u, v)$ are the velocity potential and the stream function, respectively.

In steady flow, there is flow along the boundary of the flow (the container), so the boundary of flow on the z plane corresponds to the boundary of flow on the w plane, and one of the flow lines for $\Psi =$ constant becomes that boundary. Therefore, to find the flow around a given body, it is not necessary to solve the flow in great detail in each case, because if the flow around a different body is known, it is sufficient to find a conformal transformation that maps the boundary of that body to the boundary of the other. Therefore, conformal mappings are important in connection with the irrotational flow in a two-dimensional incompressible fluid.

The same is true for problems involving electrostatic fields in two-dimensional space. For example, if a boundary is made of metal, that boundary is a surface (or a line) with constant electrostatic potential. Therefore, the electrostatic field under different boundary conditions can be known, if we can construct the conformal mapping which transforms the boundary onto another.

4.4.3 *Application to electromagnetism*

Consider a charge distributed on a line (charged line) with line density σ, extending infinitely in the z direction in three-dimensional space. Then we consider two-dimensional (x, y) space with a charged particle at $(0,0)$. The potential, in this case, is

$$\phi(x, y) = -2\sigma \ln r \qquad (r = \sqrt{x^2 + y^2}). \tag{4.29}$$

Therefore, considering the complex potential

$$\Phi = -2\sigma \log z \qquad (z = x + iy), \tag{4.30}$$

the real part ϕ of Φ is the electrostatic potential. The strength of the electric field is $|\mathrm{grad}\,\phi| = 2\sigma/r$.

Example 4.5. (Two parallel charged lines. I) Charged lines are placed at $x = \pm a$, $y = 0$, and they are parallel to the z axis in three-dimensional space, with the same line density of charge. The overall complex potential is

$$\Phi(x, y) = -2\sigma[\log(z + a) + \log(z - a)]$$
$$= -2\sigma \log(z^2 - a^2) \qquad (z = x + iy). \tag{4.31}$$

The real part ϕ of Φ is electrostatic potential, while the imaginary part gives lines of electric force (Fig. 4.7).

Example 4.6. (Two parallel charged lines. II) Let charged lines be placed at $x = \pm a$, $y = 0$, parallel to the z axis in three-dimensional space,

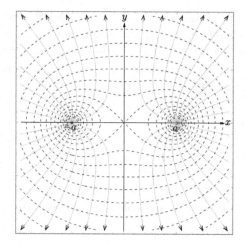

Fig. 4.7 Electric field (equipotential surface, broken lines) and lines of electric force (solid lines) when two charged lines are placed on $x = \pm a$, $y = 0$, and they are parallel to the z axis, with the same line density.

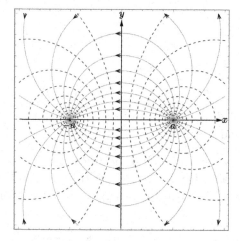

Fig. 4.8 Electric field (equipotential surface, broken lines) and lines of electric force (solid lines) when two charged lines are placed on $x = \pm a$, $y = 0$, and they are parallel to the z axis, with line densities of different signs ($\pm\sigma$).

with the line densities of opposite signs, $+\sigma$ and $-\sigma$. The overall complex potential is

$$\Phi(x, y) = -2\sigma[\log(z - a) - \log(z + a)]$$

$$= -2\sigma \log \frac{z - a}{z + a} \qquad (z = x + iy). \qquad (4.32)$$

The equipotential surfaces and lines of the electric force for the electrostatic potential can also be calculated. These are shown in Fig. 4.8.

Example 4.7. (Two parallel charged lines. III) Let us consider a slightly more complex situation. Until now, we have considered line densities with equal absolute values for the two charged lines. Assume that charged lines are placed at $x = \pm a$, $y = 0$, and they are parallel to the z axis in three-dimensional space, with the line densities of opposite signs and different densities, $+q\sigma$ and $-\sigma$. The overall complex potential is

$$\Phi(x, y) = -2\sigma[q\log(z - a) - \log(z + a)]$$

$$= -2\sigma \log \frac{(z - a)^q}{z + a} \qquad (z = x + iy). \qquad (4.33)$$

Figure 4.9 shows the equipotential surface and lines of electric force when $q = 2$.

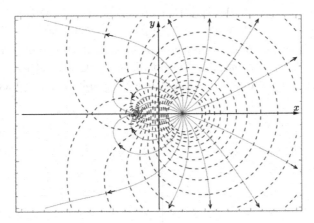

Fig. 4.9 Electric field (equipotential surface, broken lines) and lines of electric force (solid lines) when charged lines are placed on $x = \pm a$, $y = 0$, and they are parallel to the z axis, with the line densities of opposite signs and different densities, $+2\sigma$ and $-\sigma$.

Example 4.8. (Electric field around a grounded rod (plate) of zero thickness) Let us examine one example in which, using conformal mapping, equipotential curves and lines of electric force can be easily obtained.

Consider a space in which the domain $y < 0$ is the ground surface, and which extends infinitely in the z direction. Consider the electric field when a metal rod (plate) of length 1 stands parallel to the y axis (grounded) at $x = 0$. In this case, the field is uniform in the z-direction, so it is sufficient to consider only x and y.

First, let us consider the case of no metal rod. We consider a u–v space and assume that the ground surface extends at $v \leq 0$ and that there is vacuum space at $v \geq 0$. In this situation, the equipotential surface is parallel to the ground ($v = $ constant), and lines of electric force are perpendicular to the ground.

Let the u–v plane be the complex w plane ($w = u + iv$). Consider a transformation

$$w = (z^2 + 1)^{1/2}, \tag{4.34}$$

or,

$$z = (w^2 - 1)^{1/2}. \tag{4.35}$$

The real axis (u axis) on the complex w plane is mapped as follows:

$$
\begin{array}{lll}
v = 0, \ u \leq -1 & \mapsto y = 0, & x \leq 0 \\
v = 0, \ -1 \leq u \leq 0 & \mapsto 0 \leq y \leq 1, & x = 0 \\
v = 0, \ 0 \leq u \leq 1 & \mapsto 0 \leq y \leq 1, & x = 0 \\
v = 0, \ 1 \leq u & \mapsto y = 0, & 0 \leq x
\end{array}
$$

Thus, the upper half of the complex w plane is mapped to the upper half of the complex z plane containing the cut line ($x = 0$, $0 \leq y \leq 1$, corresponding to a rod). Therefore, the straight lines that follow the real axis of the

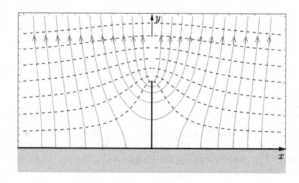

Fig. 4.10 Consider a space in which the domain of $y < 0$ is the ground surface. The electric field (equipotential surface, broken lines) and lines of electric force (solid lines) when a metal rod (plate) of length 1 stands perpendicular to the ground (parallel to the y axis) at $x = 0$.

w plane are mapped to curves that follow this cut line on the complex z plane. The equipotential lines on the w plane are a group of straight lines parallel to the u axis, while the lines of the electric force on the w plane are a group of straight lines parallel to the v axis. Resultant equipotential lines and lines of the electric force on the z plane are as shown in Fig. 4.10. This example can be regarded also as the behavior of a fluid flow along a wall.

4.4.4 *Application to fluid dynamics*

Example 4.9. (Flow around a cylinder—Joukowski's transformation) Consider the flow around a circle (cylinder) of radius a. When

$$I(C) = \oint_C \mathbf{v} \cdot d\mathbf{s} = \oint_C v_s \, ds \qquad (4.36)$$

along the closed curve C within the flow is not 0, it is called **circulation**. If, for example, an object is placed in the flow, the flow lines forms closed curves, generating a flow around the object. If the object is a circle (cylinder) and there is no circulation, the flow is symmetrical above and below the circle. If there is circulation, the flow differs between above and below the circle, so pressure works and causes lift. For simplicity, we will assume here that there is no circulation.

Consider the transformation

$$w = z + \frac{a^2}{z}, \qquad (4.37)$$

which is called **Joukowski's transformation**. Letting $z = a\,e^{i\theta}$, this becomes $w = 2a\cos\theta$, so Joukowski's transformation maps a circle of radius

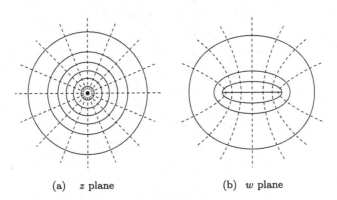

(a) z plane   (b) w plane

Fig. 4.11 Joukowski's transformation.

a on the z plane to a plate of length $4a$ on the w plane (Fig. 4.11). In the case of flow parallel to a plate of no thickness on the w plane (which is the same as the case of no plate), the complex velocity potential is

$$f = Uw. \tag{4.38}$$

In fact, if we let $w = u + iv$, then $\Phi(u, v) = Uu$, $\Psi(u, v) = Uv$, so the flow velocity in the u direction is U, while the flow velocity in the v direction is 0, and flow lines are represented as straight lines with $v = $ constant. Expressing the complex velocity potential $f = Uw$ in terms of z produces

$$f = U\left(z + \frac{a^2}{z}\right). \tag{4.39}$$

Further away from the circle, the flow ($z \to \infty$) becomes $f \to Uz$, which is a uniform flow. At the surface of a cylinder ($z = ae^{i\theta}$)

$$f = U\left(ae^{i\theta} + \frac{a^2}{ae^{i\theta}}\right) = 2Ua\cos\theta. \tag{4.40}$$

That is

$$\Phi = 2Ua\cos\theta, \qquad \Psi = 0, \tag{4.41}$$

and the flow lines certainly follow the circle (cylinder). The velocity of flow around the circle is given by

$$v_\theta = \left(\frac{1}{r}\frac{\partial\Phi}{\partial\theta}\right)\bigg|_{r=a} = -2U\sin\theta. \tag{4.42}$$

Figure 4.12 illustrates the flow.

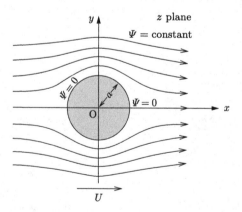

Fig. 4.12   Irrotational flow of a two-dimensional incompressible fluid around a cylinder (with no circulation).

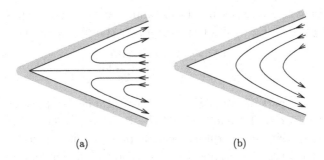

(a) (b)

Fig. 4.13 Irrotational flow around an angle π/n.

Example 4.10. (Flow that hits a flat plate diagonally) Joukowski's transformation can be used in reverse for flow striking a flat plate diagonally. A flow that strikes the circle (cylinder) at an angle α on the w plane is

$$f = U\left(e^{-i\alpha}w + \frac{a^2 e^{i\alpha}}{w}\right). \tag{4.43}$$

Therefore, as for a flow striking a flat plate $2a$ long on the z plane, at an angle of α, one can see it by transforming Eq. (4.43) using the following transformation

$$z = w + \frac{a^2}{w}.$$

Example 4.11. (Flow around the angle of a wedge) Other than Joukowski's transformation, there are a variety of applications of conformal mapping for a two-dimensional flow. Consider a case in which complex velocity potential is

$$f(z) = Az^n. \tag{4.44a}$$

The real part and imaginary part of f are $(z = re^{i\theta})$

$$\Phi = Ar^n \cos n\theta, \qquad \Psi = Ar^n \sin n\theta, \tag{4.44b}$$

so the flow line $\Psi = 0$ is $\theta = m\pi/n$ $(n = 0, 1, 2, \cdots)$. Then, the velocity potential is the flow around a wedge at an angle of $m\pi/n$ (Fig. 4.13).

4.4.4.1 *The problem of flow with circulation*

A logarithmic function is required to express circulation around a circle (cylinder). If there is circulation and flow becomes uniform at a sufficient distance from the circle, adding logarithmic terms to the complex velocity potential given by Eq. (4.39) produces a new complex potential;

$$f = U\left(z + \frac{a^2}{z}\right) + i\frac{\Gamma}{2\pi}\log z. \tag{4.45}$$

The imaginary part of the newly-added logarithmic function term is constant on the circle. Therefore, this circle $|z| = a$ becomes a flow line for complex potential. Furthermore, the complex potential increases by just $-\Gamma$ when the flow goes around the upper side of the circle $|z| = a$. This means that the flow along the circle, that is circulation flow, has been produced (Fig. 4.14).

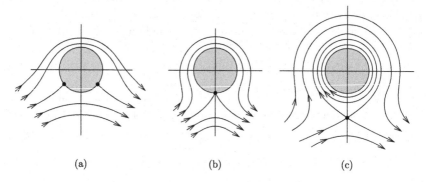

(a) (b) (c)

Fig. 4.14   Flow with circulation. (a) $\Gamma < 4\pi Ua$, (b) $\Gamma = 4\pi Ua$, (c) $\Gamma > 4\pi Ua$.

From the above velocity potential, the velocity is

$$\frac{\mathrm{d}}{\mathrm{d}z}f = U\left(1 - \frac{a^2}{z^2}\right) + i\frac{\Gamma}{2\pi}\frac{1}{z}. \tag{4.46}$$

If there is no circulation, the flow above and below the cylinder is symmetrical. On the other hand, when there is circulation, the flow is faster in the forward direction with the circulation around the cylinder and slower on the opposite side. Therefore, a force acts on the cylinder, due to Bernoulli's principle.

The velocity of flow around the cylinder is given by

$$f = \Phi + i\Psi,$$
$$v_\theta = \left(\frac{1}{r}\frac{\partial\Phi}{\partial\theta}\right)_{r=a} = -2U\sin\theta - \frac{\Gamma}{2\pi a}. \tag{4.47}$$

Letting the pressure be p_0 with no flow, and the density of the fluid be ρ, the pressure due to Bernoulli's theorem is

$$p = p_0 - \frac{\rho}{2}v_\theta{}^2 = p_0 - \frac{\rho}{2}\left(-2U\sin\theta - \frac{\Gamma}{2\pi a}\right)^2. \tag{4.48}$$

To calculate the total pressure acting on the cylinder, perform the following integral,

$$\mathbf{P} = -\int p\mathbf{n}\,ds.$$

Bearing in mind that $\mathbf{n} = (\cos\theta, \sin\theta)$ in the direction normal to the surface, and that the surface segment is $ds = a\,d\theta$. Then the pressures in the x and y directions are obtained as

$$P_x = 0, \qquad P_y = \rho U \Gamma. \tag{4.49}$$

In an incompressible perfect fluid, there is no resistance acting against the cylinder in the direction of flow (the x direction), and a lifting force acts in proportion to the circulation in the direction perpendicular to the flow (the y direction).

Chapter 5

Singularities

Knowing the properties of a complex function in the holomorphic domain also means knowing the properties of the function in the neighborhood of non-holomorphic points and non-holomorphic domains. Points for which a complex function is holomorphic are called holomorphic points, and those which are not holomorphic are called "singularities" or "singular points". The structure of Riemann surfaces is important for understanding the properties of complex functions, and this chapter is configured to reveal nature close to branch points, which are one kind of singularities.

5.1 Isolated singularity

We will start by clarifying what an isolated singularity is, and how to categorize it.

Definition 5.1. Taking R as an appropriate finite number, z_0 is called an **isolated singularity** of $f(z)$, when $f(z)$ is single-valued holomorphic for $0 < |z - z_0| < R$ and not holomorphic at $z = z_0$.

There are the following three types of isolated singularities (Secs. 5.1.1–5.1.3).

5.1.1 *Removable singularity*

Definition 5.2. When z_0 is an isolated singularity of $f(z)$, and can be re-defined as

$$f(z_0) \equiv \lim_{z \to z_0} f(z), \tag{5.1}$$

it may be possible to make $f(z)$ single-valued holomorphic in the domain that includes $z = z_0$. In such a situation, the point z_0 is called a **removable singularity**.

In general if

$$\lim_{z \to z_0} (z - z_0) f(z) = 0 \tag{5.2}$$

at the isolated singularity, then $z = z_0$ is a removable singularity of $f(z)$. This can be proved by Morera's theorem, which will be presented later, in Sec. 7.2.

Example 5.1. The equation

$$f(z) = \frac{\sin z}{z} \tag{5.3}$$

is not defined at $z = 0$, but

$$\lim_{z \to 0} f(z) = 1, \tag{5.4}$$

and $f(z)$ is bounded and single-valued holomorphic except $z = 0$. In this case, redefining $f(z)$ as

$$f(z) = \begin{cases} \sin z/z & (z \neq 0) \\ 1 & (z = 0) \end{cases}, \tag{5.5}$$

$f(z)$ is single-valued holomorphic at all points of a finite distance from the origin on the complex z plane including $z = 0$.

5.1.2 *Poles*

Definition 5.3. When z_0 is an isolated singularity of $f(z)$, and

$$\lim_{z \to z_0} |f(z)| = \infty, \tag{5.6}$$

$z = z_0$ is called a **pole**.

In the neighborhood of $z = z_0$, excluding z_0, consider a function $f(z)$;

$$f(z) = \frac{a_{-k}}{(z - z_0)^k} + \cdots + \frac{a_{-1}}{z - z_0} + a_0 + a_1(z - z_0) + \cdots \qquad (a_{-k} \neq 0), \tag{5.7}$$

where k is called the **order** of the pole $z = z_0$ of the function $f(z)$, and $z = z_0$ is called a "kth order pole" of $f(z)$. Most functions are expanded in the same way as Eq. (5.7). If an isolated singularity is a pole, the "order (of the pole)" is important.

5.1.3 *(Isolated) essential singularity*

Definition 5.4. When z_0 is an isolated singularity of $f(z)$, and is not bounded in the neighborhood of $z = z_0$, but $z = z_0$ is not a pole either, $z = z_0$ is called an (**isolated**) **essential singularity**.

As a specific example of an (isolated) essential singularity, in the area around $z = z_0$ excluding z_0 (*i.e.* for a certain R, $0 < |z - z_0| < R$), a function $f(x)$ is written as

$$f(z) = \sum_{n=-\infty}^{\infty} a_n (z - z_0)^n$$

$$= \cdots + \frac{a_{-n}}{(z - z_0)^n} + \cdots + \frac{a_{-1}}{z - z_0} + a_0 + a_1(z - z_0) + \cdots , \quad (5.8)$$

and there are infinite numbers of terms of negative power for which $a_{-n} \neq 0$ ($n > 0$). In that case, z_0 is an essential singularity. Conversely, an essential singularity can always be represented as in Eq. (5.8). The general theory is presented in Sec. 7.3.2.

If z_0 is an (isolated) essential singularity and an appropriate numerical sequence $\{z_n\}_{n=1}^{\infty}$ converges to z_0, $f(z_n)$ is arbitrary close to any arbitrary complex number, including infinity. This fact is called Weierstrass' theorem.

Theorem 5.1 (Weierstrass theorem (Casorati-Weierstrass theorem)). *If $z = z_0$ is an (isolated) essential singularity of $f(z)$, $f(z)$ is arbitrary close to any arbitrary complex number, within an arbitrarily small region $0 < |z - z_0| < \delta$ in the neighborhood of z_0.*

Proof. Let γ be any arbitrary complex number. It is sufficient to show that an appropriate positive number δ exists for any arbitrary positive number ε, such that $|f(z) - \gamma| < \varepsilon$ holds for $0 < |z - z_0| < \delta$. This may be proved by *reductio ad absurdum*. Assume that when $0 < |z - z_0| < \delta$, a complex number value β exists for which $|f(z) - \beta| \geq m$ for a given $m > 0$. In this case,

$$\phi(z) \equiv \frac{1}{f(z) - \beta} \quad (5.9)$$

satisfies that $|\phi(z)| \leq 1/m$ in a region $0 < |z - z_0| < \delta$. Thus $\phi(z)$ is holomorphic and bounded around $z = z_0$, but excluding $z = z_0$. Therefore, $\lim_{z \to z_0} \phi(z)$ converges to a finite value. Now, we can consider

$$f(z) = \beta + \frac{1}{\phi(z)} \quad (5.10)$$

separately in the following two cases.

(1) If $\phi(z) \to 0$ $(z \to z_0)$, $|f(z)|$ diverges to infinity.
(2) If $\phi(z) \to$ (a non-zero bounded value) $(z \to z_0)$, $f(z)$ converges.

In the first case, z_0 is a pole, and in the second, it is a removable singularity. This contradicts the definition of an (isolated) essential singularity. Therefore, the starting assumption is denied. $\qquad\square$

Next, we will raise a theorem that is more stringent than Weierstrass theorem without proof.

Theorem 5.2 (Picard's theorem). *If $f(z)$ is single-valued holomorphic in $0 < |z - z_0| < \rho$ and $z = z_0$ is an (isolated) essential singularity, $f(z)$ takes all finite complex numbers infinitely often with an exception of at most one value, within the domain.*

Example 5.2. We will omit the proof for Picard's theorem (sometimes call it Picard's great theorem), but examine some specific examples, to better understand Picard's theorem. Consider a complex function

$$f(z) = \exp\left(\frac{1}{z}\right). \tag{5.11}$$

In the neighborhood of $z = 0$, this is expanded into series as

$$f(z) = 1 + \frac{1}{z} + \frac{1}{z^2} + \cdots + \frac{1}{z^n} + \cdots, \tag{5.12}$$

so $z = 0$ is an (isolated) essential singularity. Let a be any arbitrary complex number value. The z for which $f(z) = a$ is

$$z_n = \frac{1}{\log a} = \frac{1}{\ln|a| + i(\arg a + 2\pi n)}. \tag{5.13}$$

Thus, $f(z)$ takes the value a in the point sequence $\{z_n\}_{n=1}^{\infty}$. When $n \to \infty$, $z_n \to 0$. However, $f(z)$ cannot take a value 0. This is the above-mentioned exception. Therefore, $\exp(1/z)$ can take any arbitrary complex number in the neighborhood of $z = 0$, except 0, infinite times.

Example 5.3. For e^z, $\sin z$, $z = \infty$ is an (isolated) essential singularity. For any other arbitrary point, which is to say, for any arbitrary point of a finite distance from the origin on the complex plane, these functions are holomorphic.

5.2 Accumulation singularity

In addition to the (isolated) essential singularity described in Sec. 5.1.3, there is another essential singularity. For example, a function

$$f(z) = \operatorname{cosec} \frac{1}{z} = \frac{1}{\sin \frac{1}{z}} \qquad (5.14)$$

has singularities at $z = 1/n\pi$ ($n = \pm 1, \pm 2, \cdots$), which are first-order poles if n is finite. There are an infinite number of these singular points (poles) in the neighborhood of $z = 0$. We cannot imagine a circle centered at $z = 0$ and containing no singular point. This kind of point $z = 0$ is called an **accumulation singularity**, and it is a kind of essential singularities. An expansion such as that in Eq. (5.8) is not possible in the neighborhood of an accumulation singularity.

5.3 Branch points

In a function $w = f(z)$ such as nth root and logarithmic function, we see a point z_0 for which, when z rotates around z_0 by 2π and returns to the original point, w does not return to the original value. This kind of point z_0 is called a **branch point**.

Example 5.4. Consider $w = f(z) = z^{1/2}$. Let z move on the unit circle from the start point $z = 1$ on the complex z plane, as shown in Fig. 5.1. The function $z^{1/2}$ is a two-valued function, and we can assign 0 or 2π as arguments of $z = 1$. Then, $1^{1/2}$ is either 1 (argument 0) or -1 (argument π).

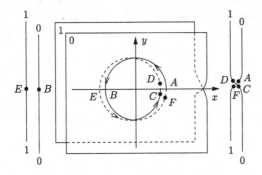

Fig. 5.1   Branch cut for $z^{1/2}$, and Riemann surfaces.

(1) First, determine that the argument of $z^{1/2}$ is 0 when $z = 1$ (point A). This means that we have chosen $\arg z = 0$ when $z = 1$. Write $z = e^{i\theta}$ and raise θ from 0.

(2) When z starts from the argument 0 and rotates around the origin by π, $\arg z = \pi$ (point B), $(-1)^{1/2} = (e^{i\pi})^{1/2} = e^{i\pi/2} = i$.

(3) Furthermore, when z moves around the origin by π, it reaches $\arg z = 2\pi$, $z = 1$ (point C), and $(1)^{1/2} = (e^{i2\pi})^{1/2} = e^{i\pi} = -1$. Here, $z = 1$, but it linked to $(1)^{1/2} = -1$ (point D).

(4) A further rotation around the origin produces $\arg z = 4\pi$, $z = 1$ (point F), and $(1)^{1/2} = (e^{i4\pi})^{1/2} = e^{i2\pi} = 1$. This means that z returns to the initial point (point F = point A) and w returns to the initial value.

In the above example, the value of $z^{1/2}$ returns to its original value after twice circular rotation around $z = 0$. The argument of z changed in this case from 0 to 4π, but even if the point $0 < \arg z \leq 2\pi$ and the point $2\pi < \arg z \leq 4\pi$ are the same on the z plane, the function value $f(z) = z^{1/2}$ is different. Considering these two sets of z planes as different ones, and considering the z plane corresponding to $0 < \arg z \leq 2\pi$ and the z plane corresponding to $2\pi < \arg z \leq 4\pi$ as two separate sheets of z plane is the concept of Riemann surface. For the function $z^{1/2}$, we need two z planes, so this is called a two–sheet Riemann surface.

The branch points for $z^{1/2}$ are $z = 0$ and $z = \infty$. On two z planes, **cut** between the two branch points $z = 0$ and $z = \infty$, inserting a branch cut. Here the right side of the real axis is branch cut, but any other kind of curve that starts from $z = 0$ and reaches $z = \infty$ is acceptable. The value of the function is the same on the branch cut with $\arg z = 2\pi$ on the first and the second z planes, so they can be glued together here. Furthermore, the value of the function is the same on the branch cut with $\arg z = 4\pi$ on the second z plane and that with $\arg z = 0$ on the first z plane, so they can be glued together here. This produces Riemann surface with two sheets which associates z with $z^{1/2}$ on a one-to-one basis.

Example 5.5. The situation is even more complicated in cases where there are more branch points. Consider two functions

$$w = (z - 1)^{1/2}(z - 2)^{1/2} \tag{5.15}$$

and

$$w = (z - 1)^{1/2}(z - 2)^{1/3}. \tag{5.16}$$

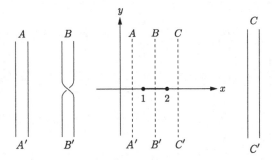

Fig. 5.2   Riemann surface for $w = (z-1)^{1/2}(z-2)^{1/2}$. The branch points are $z = 1$ and $z = 2$, and the line to link $z = 1$ and $z = 2$ is chosen as the branch cut.

In the first case, $z = \infty$ is not a branch point for $w = (z-1)^{1/2}(z-2)^{1/2}$, but only $z = 1$ and $z = 2$ are branch points. That is because when the transformation $z = 1/\zeta$ is used,

$$w = \left(\frac{1}{\zeta} - 1\right)^{1/2} \left(\frac{1}{\zeta} - 2\right)^{1/2} = \frac{1}{\zeta}(1 - \zeta)^{1/2}(1 - 2\zeta)^{1/2}$$

and $\zeta = 0$ ($z = \infty$) is a pole, not a branch point. Therefore, the branch cut should choose the line to link $z = 1$ and $z = 2$. Figure 5.2 shows Riemann surface.

In the second case, the branch points for $w = (z-1)^{1/2}(z-2)^{1/3}$ are $z = 1, 2, \infty$. Therefore, two lines are chosen for the branch cut: the half line joining $z = 1$ and $z = \infty$ (the half line starting from $z = 1$ and extending to the left) and the half line joining $z = 2$ and $z = \infty$ (the half line starting from $z = 2$ and extending to the right). Circular rotation of z twice around $z = 1$ and that three times around $z = 2$ makes the value of w return to the original value. Circular rotation six times around $z = \infty$ is required for w to return to the original value. Let us consider this more directly. Performing a variable transformation with $z = 1/\zeta$ produces

$$w = \left(\frac{1}{\zeta} - 1\right)^{1/2} \left(\frac{1}{\zeta} - 2\right)^{1/3} = \frac{1}{\zeta^{5/6}}(1 - \zeta)^{1/2}(1 - 2\zeta)^{1/3}.$$

This includes the term $\zeta^{5/6}$, so $\zeta = 0$ ($z = \infty$) is a branch point, and circular rotation six times around $\zeta = 0$ brings the argument of the function w to its original value for the first time. Therefore, six z planes are required, and they are linked at the branch cut. Figure 5.3 shows Riemann surface of six sheets.

Comparing Figs. 5.2 and 5.3, the forms of the functions appear similarly in the first and second cases, but their branch points differ, and the way

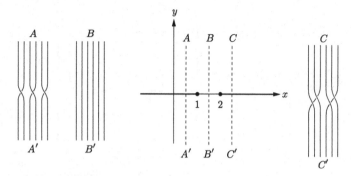

Fig. 5.3   Riemann surface for $w = (z-1)^{1/2}(z-2)^{1/3}$. The branch points are $z = 1$, 2, ∞, and the half line to link $z = 1$ and $z = \infty$ and the half line to link $z = 2$ and $z = \infty$ are chosen as the branch cut.

of cutting the branch is different. Therefore, it can be understood that the structure of Riemann surface is different.

When the index of the power function is a rational number, the number of z planes forming Riemann surface is finite. In that case, the branch point is called an **algebraic branch point**. When the power index is not a rational number (generalized exponential function), or for the logarithmic function $w = \log z$, it is an infinitely multi-valued function, and an infinite number of z planes form Riemann surface. This kind of branch point is called a **logarithmic branch point**. In the case of an algebraic branch point, the path of z can be closed by the number of circular rotations in the same direction around that point. In the case of a logarithmic branch point, however, to close a path on Riemann surface, not only one-way circular rotation but the same number of opposite-direction circular rotations are required as well.

Chapter 6

Complex Integrals

In this chapter, the integral of a complex function is defined by using the line integrals, and its fruitful properties are examined. These are based on Cauchy's integral theorem and the following residue theorem. Employing the residue theorem makes it easy to find the definite integral in many cases. Complex integral opens up a much deeper and broader world of complex functions.

6.1 Closed Jordan curves and the shape of the holomorphic domain

Taking a real variable t as a parameter, suppose that the point $z = z(t)$ moves continuously on a complex z plane and draws a single curve. Take $z(a)$ as the initial point and $z(b)$ as the final point, where $a \leq t \leq b$. The curve is called a closed curve when $z(a) = z(b)$. If the curve does not intersect with itself, which is to say, $z(t_1) \neq z(t_2)$ $(t_1 \neq t_2)$ except at the

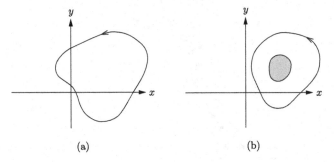

(a)   (b)

Fig. 6.1   Closed Jordan curve and domain. (a) Simply-connected domain, (b) multiply-connected domain. In (b), the shaded part is surrounded by the domain D.

initial and final points, such a curve is called **Jordan curve**. Jordan curve with $z(a) = z(b)$ is called a **closed Jordan curve**.

A complex plane is divided by a closed Jordan curve into two parts. The bounded domain is called the **interior**, and the unbounded remainder is called the **exterior**. On the closed Jordan curve, the forward direction in which the interior appears on the left is the **positive direction** (counter-clockwise direction), and the opposite direction is the **negative direction** (clockwise direction). In domain D, if it is possible to continuously deform a closed Jordan curve into a single point, that domain D is said to be **simply-connected**. A domain that is not simply-connected is said to be **multiply connected** (Fig. 6.1). If there is no possibility of confusion, a closed Jordan curve is often referred to simply as a closed curve.

6.2 Definition of Complex Integral

Let us define the integral of a complex function (complex integral) on a specific path on a complex plane. A continuous complex function $f(z)$ is defined in a domain D, and there is a continuous path C in D. Continuous curve C consists of a smooth curve or a finite number of joins thereof. This kind of curve is called a **piecewise smooth curve**. In the following, each path is taken to be piecewise smooth.

Let the initial point of C be z_0, and the final point be z. Add division points $z_1, z_2, \cdots, z_{N-1}$ on C between z_0 and z, and express that division as

$$\Delta = \{z_0, z_1, z_2, \cdots, z_{N-1}, z_N = z\}. \tag{6.1}$$

On the division Δ, take an arbitrary point ζ_j between z_{j-1} and z_j, and

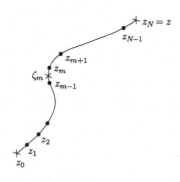

Fig. 6.2   Division Δ on Curve C.

consider the finite sum (Fig. 6.2)

$$S_\Delta = \sum_{j=1}^{N} f(\zeta_j)(z_j - z_{j-1}). \tag{6.2}$$

Definition 6.1. When the number of the division points of division Δ is infinite, and the length of the interval $|z_j - z_{j-1}|$ is infinitesimally small, the limit value of the sum S_Δ for the continuous function $f(z)$ is assumed to be finite and uniquely determined. In this situation, it is said that "$f(z)$ is complex integrable", and the value is called the **complex integral**. This complex integral is written as

$$\int_C f(z)\,\mathrm{d}z = \lim_{\delta \to 0} \sum_{j=1}^{N} f(\zeta_j)(z_j - z_{j-1}). \tag{6.3}$$

Here, $\delta = \max |z_j - z_{j-1}|$ and, as $\delta \to 0$, the number N of divisions becomes infinite. Curve C, including its initial and final points and direction, is called the **path of integration**. When the path of integration C is a closed Jordan curve which is called **contour (contour path)**,[1] the direction of the integration is in the positive direction or negative direction, according to the direction of the path. When the path of integration C is **closed** and the point z moves in the positive direction on a closed Jordan curve, this is written as

$$\oint_C f(z)\,\mathrm{d}z. \tag{6.4}$$

We will now demonstrate that if the complex function $f(z)$ is continuous, the limit value of the sum of S_Δ, when $\delta = \max |z_j - z_{j-1}| \to 0$, is a finite and determinate value. Write the real parts and imaginary parts of z_j, ζ_j and $f(\zeta_j)$ as

$$z_j = x_j + \mathrm{i}y_j, \qquad \zeta_j = \xi_j + \mathrm{i}\eta_j, \qquad f(\zeta_j) = u_j + \mathrm{i}v_j.$$

S_Δ is rewritten to obtain

$$\begin{aligned}
S_\Delta &= \sum_{j=1}^{N} f(\zeta_j)(z_j - z_{j-1}) \\
&= \sum_{j=1}^{N} \{[u_j(x_j - x_{j-1}) - v_j(y_j - y_{j-1})] + \mathrm{i}[v_j(x_j - x_{j-1}) + u_j(y_j - y_{j-1})]\}.
\end{aligned} \tag{6.5}$$

---

[1] It should be noted that, in several textbooks, the word "contour" is used synonymously with the word "path".

Using a parameter s, the path of integration C is expressed as

$$C : x = x(s), \qquad y = y(s) \qquad (a \le s \le b). \tag{6.6}$$

Then we get

$$z_j - z_{j-1} = z(s_j) - z(s_{j-1}) \approx \frac{\mathrm{d}z}{\mathrm{d}s}(s_j - s_{j-1}),$$

$$x_j - x_{j-1} = x(s_j) - x(s_{j-1}) \approx \frac{\mathrm{d}x}{\mathrm{d}s}(s_j - s_{j-1}),$$

$$y_j - y_{j-1} = y(s_j) - y(s_{j-1}) \approx \frac{\mathrm{d}y}{\mathrm{d}s}(s_j - s_{j-1})$$

so the sum is rewritten as

$$S_\Delta = \sum_j \left(u_j \frac{\mathrm{d}x}{\mathrm{d}s} - v_j \frac{\mathrm{d}y}{\mathrm{d}s} \right) (s_j - s_{j-1}) + \mathrm{i} \sum_j \left(v_j \frac{\mathrm{d}x}{\mathrm{d}s} + u_j \frac{\mathrm{d}y}{\mathrm{d}s} \right) (s_j - s_{j-1}). \tag{6.7}$$

Two functions of s

$$u(x(s), y(s)) \frac{\mathrm{d}x}{\mathrm{d}s} - v(x(s), y(s)) \frac{\mathrm{d}y}{\mathrm{d}s},$$

$$v(x(s), y(s)) \frac{\mathrm{d}x}{\mathrm{d}s} + u(x(s), y(s)) \frac{\mathrm{d}y}{\mathrm{d}s} \tag{6.8}$$

are piecewise continuous functions, so S_Δ in Eq. (6.7) is confirmed as a finite value in limit $\delta \to 0$, $|s_j - s_{j-1}| \to 0$ (see [T. Takagi (1961)] concerning "Riemann integral theorem"). This is written as

$$\lim S_\Delta = \int_a^b \left(u \frac{\mathrm{d}x}{\mathrm{d}s} - v \frac{\mathrm{d}y}{\mathrm{d}s} \right) \mathrm{d}s + \mathrm{i} \int_a^b \left(v \frac{\mathrm{d}x}{\mathrm{d}s} + u \frac{\mathrm{d}y}{\mathrm{d}s} \right) \mathrm{d}s \tag{6.9}$$

and is called the line integral along curve C. This is also written as

$$\lim S_\Delta = \int_C (u \, \mathrm{d}x - v \, \mathrm{d}y) + \mathrm{i} \int_C (v \, \mathrm{d}x + u \, \mathrm{d}y) = \int_C f(z) \, \mathrm{d}z. \tag{6.10}$$

This is the complex integral along the path of integration C on a complex plane.

Example 6.1. Let us try integrating $f(z) = z$ along three paths of integration that have the initial point $z = 0$ and the final point $z = 1 + \mathrm{i}$, as shown in Fig. 6.3, according to the definition of complex integral.

(1) $C_1 : 0 \to 1 \to 1 + \mathrm{i}$.
(2) $C_2 : 0 \to 1 + \mathrm{i}$.
(3) $C_3 : 0 \to \sqrt{2} \xrightarrow{\mathrm{arc}} 1 + \mathrm{i}$.

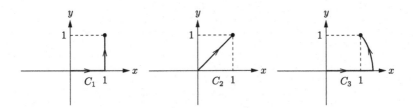

Fig. 6.3 Paths of integration C_1, C_2, C_3.

(1) On C_1, $z = x$ in the part $0 \to 1$, so $\mathrm{d}z = \mathrm{d}x$. In the part $1 \to 1 + \mathrm{i}$, $z = 1 + \mathrm{i}y$, so $\mathrm{d}z = \mathrm{i}\,\mathrm{d}y$. Then

$$\int_{C_1} z\,\mathrm{d}z = \int_0^1 x\mathrm{d}x + \mathrm{i}\int_0^1 (1 + \mathrm{i}y)\,\mathrm{d}y = \frac{1}{2} + \mathrm{i}\left(1 + \mathrm{i}\frac{1}{2}\right) = \mathrm{i}.$$

(2) On C_2, $z = (1 + \mathrm{i})\,s$ and $\mathrm{d}z = (1 + \mathrm{i})\,\mathrm{d}s$. Then

$$\int_{C_2} z\,\mathrm{d}z = \int_0^1 (1 + \mathrm{i})\,s(1 + \mathrm{i})\,\mathrm{d}s = (1 + \mathrm{i})^2 \frac{1}{2} = \mathrm{i}.$$

(3) On C_3, in the part $0 \to \sqrt{2}$, $z = x$ so $\mathrm{d}z = \mathrm{d}x$. On the arc $\sqrt{2} \to 1 + \mathrm{i}$, $z = \sqrt{2}\mathrm{e}^{\mathrm{i}\theta}$ so $\mathrm{d}z = \mathrm{i}\sqrt{2}\mathrm{e}^{\mathrm{i}\theta}\mathrm{d}\theta$. Then

$$\int_{C_3} z\,\mathrm{d}z = \int_0^{\sqrt{2}} x\,\mathrm{d}x + \int_0^{\pi/4} \sqrt{2}\,\mathrm{e}^{\mathrm{i}\theta}\,\mathrm{i}\sqrt{2}\,\mathrm{e}^{\mathrm{i}\theta}\,\mathrm{d}\theta = 1 + (\mathrm{i} - 1) = \mathrm{i}.$$

The answer is i on any of these paths of integration.

Remark 6.1. The integral of a holomorphic function in the simply-connected domain is determined by the choice of the initial and final points only, independent of the path of integration. The details are described in Sec. 6.4.2.

Let us consider one other example.

Example 6.2. Choosing $z = 1$ as the initial point and $z = 1$ as the final point, integrate $f(z) = 1/z$ along a contour path that is traversed on the unit circle in the positive direction (Fig. 6.4). On the unit circle, taking $z = \mathrm{e}^{\mathrm{i}\theta}$, $\mathrm{d}z = \mathrm{i}\mathrm{e}^{\mathrm{i}\theta}\mathrm{d}\theta$, so

$$\oint_{|z|=1} \frac{1}{z}\,\mathrm{d}z = \int_0^{2\pi} \frac{1}{\mathrm{e}^{\mathrm{i}\theta}}\mathrm{i}\mathrm{e}^{\mathrm{i}\theta}\mathrm{d}\theta = \mathrm{i}\int_0^{2\pi} \mathrm{d}\theta = 2\pi\mathrm{i}. \tag{6.11}$$

Here we indicate that the contour path is a closed curve by using the symbol $\oint$. Note that on a complex plane, the integration of 2π-round around the pole ($z = 0$) gives a non-zero value.

Next, the point z starts from $z = \mathrm{e}^{\mathrm{i}\theta_0}$ and moves to $z = \mathrm{e}^{\mathrm{i}\theta_1}$ (or $\mathrm{e}^{-\mathrm{i}(2\pi - \theta_1)}$) in the positive (or negative) direction on a unit circle (then

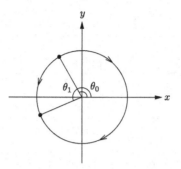

Fig. 6.4   Integral on a unit circle.

$z = e^{i\theta_1} = e^{-i(2\pi-\theta_1)}$). Since $z = e^{i\theta}$ and $dz = iz\,d\theta$, so

$$\int_{\theta=\theta_0}^{\theta=\theta_1} \frac{1}{z}\,dz = i\int_{\theta_0}^{\theta_1} d\theta = i(\theta_1 - \theta_0), \qquad (6.12a)$$

$$\int_{\theta=\theta_0}^{\theta=-(2\pi-\theta_1)} \frac{1}{z}\,dz = i\int_{\theta_0}^{-2\pi+\theta_1} d\theta = i(\theta_1 - \theta_0 - 2\pi). \qquad (6.12b)$$

In this example, the initial point and final point are the same on the z plane, but the paths of integration are different, and so are the integral values. Considering a unit circle around the pole $z = 0$, the first integral on that circle is in the positive direction and the second integral is in the negative direction. Running on the first path, and then running in reverse on the second path, results in one round rotation on the unit circle.

$$\int_{\theta=\theta_0}^{\theta=\theta_1} \frac{1}{z}\,dz - \int_{\theta=\theta_0}^{\theta=-(2\pi-\theta_1)} \frac{1}{z}\,dz = \oint_{|z|=1} \frac{1}{z}\,dz = 2\pi i. \qquad (6.12c)$$

Let us consider the cases in which, as in Example 6.1, (the complex integral from A to B on the plane z), and assume that the complex integral is determined solely by the points A and B, not by the path of integration. Consider two paths of integration C_1 and C_2, which go from A to B without

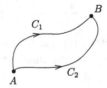

Fig. 6.5   Two paths of integration, C_1 and C_2, from point A to point B.

intersecting (Fig. 6.5). Since

$$\int_{A(C_1)}^{B} f(z)\,dz = \int_{A(C_2)}^{B} f(z)\,dz, \tag{6.13a}$$

then

$$\int_{A(C_1)}^{B} f(z)\,dz - \int_{A(C_2)}^{B} f(z)\,dz = \left(\int_{A(C_1)}^{B} + \int_{B(-C_2)}^{A} \right) f(z)\,dz = 0.$$

$$\tag{6.13b}$$

Here, the path running backward from B to A is written as $-C_2$. A minus symbol is added, reversing the path, according to the definition of an integral (6.3). The path running from A to B along C_1 and then returning from B to A $(-C_2)$ is denoted by C, and C is a closed contour path, so formula (6.13b) is rewritten as

$$\oint_C f(z)\,dz = 0. \tag{6.14}$$

The following conclusion can be derived from the above discussion.

Theorem 6.1. *The fact that a complex integral in a given domain is determined solely by the initial point and final point means that the single contour integration, taking any arbitrary closed Jordan curve in the domain as the contour path, is zero.*

6.3 Basic property of complex integral

The basic property of complex integral is summarized below. Suppose that $f(z)$ and $g(z)$ are continuous functions and that each path of integration is piecewise smooth.

(1) From the definition, linearity is satisfied for the integral. Thus, taking a as a complex constant,

$$\int_C [f(z) \pm g(z)]\,dz = \int_C f(z)\,dz \pm \int_C g(z)\,dz, \tag{6.15a}$$

$$\int_C af(z)\,dz = a \int_C f(z)\,dz. \tag{6.15b}$$

(2) When the path of integration from point A to point B is path C, the path from B to A on the same curve is written as $-C$. The formula below is satisfied for integral along paths of integration C and $-C$.

$$\int_{-C} f(z)\,dz = - \int_C f(z)\,dz. \tag{6.16}$$

Thus, as summarized in Eq. (6.13b), a minus symbol is added to the value of the complex integral when the path is reversed.

(3) The smooth path of integration C is written as $z = z(s)$ $(a \leq s \leq b)$, using the parameter s. In that case,

$$\left| \int_C f(z)\, dz \right| \leq \int_a^b |f(z(s))||z'(s)|\, ds \equiv \int_C |f(z)||dz| \qquad (6.17)$$

is satisfied.

The properties (1) and (2) have already been explained. We will now prove the property (3), *i.e.* the inequality Eq. (6.17).

Proof. The left-hand side of the inequality Eq. (6.17) is

$$\left| \lim \sum_j f(z(t_j)) \frac{dz(t_j)}{ds}(s_j - s_{j-1}) \right|, \qquad (6.18a)$$

and the right-hand side is

$$\lim \sum_j |f(z(t_j))| \left| \frac{dz(t_j)}{ds} \right| (s_j - s_{j-1}), \qquad (6.18b)$$

where, $s_{j-1} < t_j < s_j$. The inequality equation for the complex number a_j

$$\left| \sum_j a_j \right| \leq \sum_j |a_j|$$

is satisfied. Applying this to Eqs. (6.18a) and (6.18b), the inequality Eq. (6.17) is obtained. □

6.4 Cauchy's integral theorem

Cauchy's integral theorem described below is derived solely from the single-valued holomorphy of $f(z)$, that is to say, the differentiability of f (the existence of u_x, u_y, v_x and v_y for $f = u + iv$). Based on this theorem, various properties of integrals of complex functions are derived.

6.4.1 Cauchy's integral theorem

Theorem 6.2 (Cauchy's integral theorem). *In the simply-connected domain D, $f(z)$ is single-valued holomorphic, and the closed Jordan curve C lies in the domain D. In that case, the equation*

$$\oint_C f(z)\, dz = 0 \qquad (6.19)$$

holds.

Other than the holomorphy (differentiability) of $f(z) = u + iv$, we will prove this theorem, for simplicity, starting with an assumption of the continuity of the partial differential coefficients u_x, u_y, v_x and v_y. On that assumption, it is possible to prove the integral theorem using Green's theorem (a proof by Cauchy himself).

Remark 6.2. (Green's theorem) Let $f(x, y)$ and f_x, f_y be continuous in a two-dimensional domain. In this situation, taking a closed contour path in the domain as C, and the interior to the path as D, it follows that

$$\iint_D f_y(x, y)\, dx\, dy = -\oint_C f(x, y)\, dx,$$

$$\iint_D f_x(x, y)\, dx\, dy = \oint_C f(x, y)\, dy.$$

If the contour path C is expressed as $y = y_1(x)$; $a \leq x \leq b$ and $y = y_2(x)$; $a \leq x \leq b$, and always $y_1(x) \leq y_2(x)$ (Fig. 6.6), it is easy to prove the first equation. If not, the domain D should be divided into small subdomains which satisfy the above conditions.

From the above conditions, the integral is rewritten in the following way, and one can obtain the first equation.

$$\iint_D f_y\, dx\, dy = \int_a^b dx \int_{y_1}^{y_2} f_y\, dy$$

$$= \int_a^b dx\, [f(x, y_2(x)) - f(x, y_1(x))] = -\oint_C f(x, y)\, dx.$$

It is important to note the sign here. Also, if the contour path C is expressed as $x = x_1(y)$, $c \leq y \leq d$ and $x = x_2(y)$, $c \leq y \leq d$, and $x_1(y) \leq x_2(y)$, the

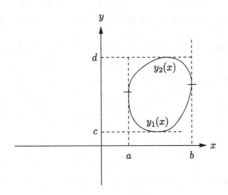

Fig. 6.6   Green's theorem.

second equation can be obtained.

$$\iint_D f_x \, \mathrm{d}x \, \mathrm{d}y = \int_c^d \mathrm{d}y [f(x_2(y), y) - f(x_1(y), y)] = \oint_C f(x, y) \, \mathrm{d}y.$$

From the two equations above, we obtain

$$\iint_D \left(\frac{\partial g}{\partial x} - \frac{\partial f}{\partial y} \right) \mathrm{d}x \, \mathrm{d}y = \oint_C (f \, \mathrm{d}x + g \, \mathrm{d}y).$$

This is **Green's theorem**.

Proof. (Proof of Cauchy's integral theorem using Green's theorem) Rewrite Eq. (6.10) using Green's theorem.

$$\oint f(z) \, \mathrm{d}z = \oint (u \, \mathrm{d}x - v \, \mathrm{d}y) + \mathrm{i} \oint (v \, \mathrm{d}x + u \, \mathrm{d}y)$$

$$= \int_D (-v_x - u_y) \, \mathrm{d}x \, \mathrm{d}y + \mathrm{i} \int_D (u_x - v_y) \, \mathrm{d}x \, \mathrm{d}y. \qquad (6.20)$$

Using the Cauchy–Riemann relation here, $u_x = v_y$, $v_x = -u_y$ [Eq. (2.16)], the integrands in the final equation of Eq. (6.20) become identically 0. This can be summarized to produce the equation

$$\oint f(z) \, \mathrm{d}z = 0.$$

$\square$

The above is the proof of Cauchy's integral theorem, using Green's theorem. We will later derive the infinitely continuous differentiability of $f(z)$ from Cauchy's integral theorem. If we assume the continuity of u_x etc. in the proof of Cauchy's integral theorem as above, it becomes circular reasoning (*circulus in probando*). To avoid that, we assume only the existence of u_x, u_y, v_x and v_y, and present proof that does not assume continuity.

Proof. (Proof of Cauchy's integral theorem without assumption of continuity of u_x etc.) For simplicity, we take the paths of integration C as a closed triangle. If it is a general curve, it may be divided into triangles. The holomorphic function $f(z)$ is differentiable, so it can be expanded around any arbitrary point z_0 in the domain, and expressed as

$$f(z) = f(z_0) + f'(z_0)(z - z_0) + \gamma. \qquad (6.21\mathrm{a})$$

Taking $|z - z_0| \to 0$, γ is a complex number that becomes 0 faster. Then, choosing a suitable positive number δ for an arbitrary positive number ε,

$$|f(z) - [f(z_0) + f'(z_0)(z - z_0)]| < \varepsilon |z - z_0| \qquad (6.21\mathrm{b})$$

for all values of z satisfying $|z - z_0| < \delta$. Furthermore, choosing z as $|z - z_0| \to 0$, one can always have $\varepsilon \to 0$. From property (3) of a complex integral, [Eq. (6.17)], along a closed Jordan curve C' that bounds a small domain including z_0 within it,

$$\left| \oint_{C'} dz\{f(z) - [f(z_0) + f'(z_0)(z - z_0)]\} \right| < \varepsilon \oint_{C'} |dz||z - z_0| \qquad (6.21c)$$

is true. As Example 6.1 shows, the integral of the linear equation does not depend on the contour path, so the result of performing a contour integral on the content of [ ] on the left-hand side is

$$\oint_{C'} [f(z_0) + f'(z_0)(z - z_0)]\, dz = 0.$$

From that,

$$\left| \oint_{C'} f(z)\, dz \right| < \varepsilon \int_{C'} |z - z_0||dz| \qquad (6.21d)$$

is obtained. Taking the length of curve C' that bounds a small sub-domain to be l, it follows that $|z - z_0| < l$, so

$$\left| \oint_{C'} f(z)\, dz \right| < \varepsilon l^2. \qquad (6.21e)$$

Joining the midpoints of each of the sides of the triangular contour path C divides it into four triangles of equal area ($C_1^{(1)}$, $C_1^{(2)}$, $C_1^{(3)}$, and $C_1^{(4)}$) (Fig. 6.7). The integral along C is equal to the sum of the integrals along each of the four triangles.

$$\oint_C = \oint_{C_1^{(1)}} + \oint_{C_1^{(2)}} + \oint_{C_1^{(3)}} + \oint_{C_1^{(4)}}. \qquad (6.21f)$$

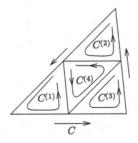

Fig. 6.7   Division of the triangular contour path C.

Taking the triangle for which the absolute value of the integral is the maximum to be $C_1^{(1)}$,

$$\left| \oint_C f(z)\,\mathrm{d}z \right| \le \sum_j \left| \oint_{C_1^{(j)}} f(z)\,\mathrm{d}z \right| \le 4 \left| \oint_{C_1^{(1)}} f(z)\,\mathrm{d}z \right|. \tag{6.21g}$$

Performing this procedure N times, one can obtain

$$\left| \oint_C f(z)\,\mathrm{d}z \right| \le 4^N \left| \oint_{C_N^{(1)}} f(z)\,\mathrm{d}z \right|. \tag{6.21h}$$

Considering Eq. (6.21e), let the small triangle C' be $C_N^{(1)}$. In that case, if we take the perimeter length of C to be L, the perimeter length of $C_N^{(1)}$ is $l = 2^{-N}L$, so

$$\left| \oint_C f(z)\mathrm{d}z \right| \le 4^N \cdot \varepsilon 4^{-N} L^2 = \varepsilon L^2. \tag{6.21i}$$

Since $\varepsilon \to 0$ can be produced by making $N \to \infty$, it follows that $|\oint_C f(z)\,\mathrm{d}z| \to 0$ and, therefore,

$$\oint f(z)\,\mathrm{d}z = 0.$$

$\square$

In the proof of Cauchy's integral theorem provided here, we should emphasize that we employed nothing other than continuity and differentiability of the function $f(z)$ and the division of the contour path C.

In previous considerations of Cauchy's integral theorem, the holomorphic domain was assumed to be simply connected. When the domain is doubly-connected, consider two contours, C_1 and C_2, which contain a non-holomorphic domain D' as shown in Fig. 6.8. Also consider the path of integration C_0 which connects C_1 to C_2, and C_0' $(-C_0)$ following in the

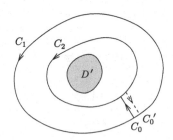

Fig. 6.8　The doubly-connected domain D and paths of integration C_1, C_2.

reverse direction to C_0. The path that runs backward on the inner path C_2 is written as $-C_2$. The contour path that joins C_1, C_0, $-C_2$, and C_0' is a closed curve that is traversed in the holomorphic domain in the positive direction only, and its inside is single-valued holomorphic. Therefore,

$$\left(\int_{C_1} + \int_{C_0} + \int_{-C_2} + \int_{C_0'} \right) f(z)\, dz = \oint f(z)\, dz = 0. \qquad (6.22)$$

The function $f(z)$ is single-valued holomorphic at each point interior to and on the contour path, so $f(z)$ on C_0 and $f(z)$ on C_0' are equal. C_0 and C_0' have opposite integration directions, so

$$\int_{C_0} f(z)\, dz + \int_{C_0'} f(z)\, dz = 0. \qquad (6.23a)$$

Also, from Eq. (6.16), it follows that

$$\int_{-C_2} f(z)\, dz = - \int_{C_2} f(z)\, dz. \qquad (6.23b)$$

Therefore,

$$\int_{C_1} f(z)\, dz = \int_{C_2} f(z)\, dz. \qquad (6.23c)$$

The above can be combined to produce the following theorem.

Theorem 6.3. *$f(z)$ is holomorphic in the domain D, but a domain D' is surrounded by D and $f(z)$ is not holomorphic in D' (that is to say, D is a doubly-connected domain). Take two closed Jordan curves in which D' is inside (Fig. 6.8). Let C_1 be possible to be continuously deformed in domain D into C_2. In that case,*

$$\oint_{C_1} f(z)\, dz = \oint_{C_2} f(z)\, dz \qquad (6.24)$$

is satisfied.

Assume a domain D where the function $f(z)$ is defined to be a single-valued holomorphic to be multiply connected (Fig. 6.9). In this situation, Cauchy's integral theorem can be generalized as follows:

Theorem 6.4. *The single-valued holomorphic domain D of $f(z)$ is multiply connected, and the non-holomorphic domains surrounded by D are $D_1, D_2, \cdots$ (Fig. 6.9). Let C_j be a closed integration path in the positive direction, with D_j interior to the path and no other non-holomorphic domains. Also, take C to be the contour path that turns in the positive direction and that contains all C_j inside it. In that case, we get*

$$\oint_C f(z)\, dz = \sum_j \oint_{C_j} f(z)\, dz. \qquad (6.25)$$

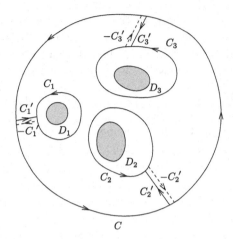

Fig. 6.9   Multiply connected domain D and closed contour path C_j.

Proof. As shown in Fig. 6.9, add a returning path C_j' between contour paths C and C_j. Provided C_j' connects C and C_j, it may be added in any way. If the path moving in reverse along C_j is written $-C_j$, a closed contour path that includes no domain of non-holomorphy within it can be created from C, $-C_j$, C_j' and $-C_j'$. If Cauchy's integral theorem is applied, one gets

$$\left[\oint_C + \sum_j \left(\oint_{-C_j} + \int_{C_j'} + \int_{-C_j'} \right) \right] f(z)\,\mathrm{d}z = 0. \tag{6.26a}$$

Further, $f(z)$ is equal on C_j' and on $-C_j'$ and

$$\left(\int_{C_j'} + \int_{-C_j'} \right) f(z)\,\mathrm{d}z = 0. \tag{6.26b}$$

Therefore,

$$\oint_C f(z)\,\mathrm{d}z = -\sum_j \oint_{-C_j} f(z)\,\mathrm{d}z = \sum_j \oint_{C_j} f(z)\,\mathrm{d}z \tag{6.26c}$$

is obtained.   □

6.4.2   *Indefinite integral and its holomorphy*

From the discussion so far, the integral $\displaystyle\int_a^z f(\xi)\,\mathrm{d}\xi$ in a simply-connected holomorphic domain does not depend on the path of integration from a to

z, and can be written as

$$F(z) = \int_a^z f(\xi)\,\mathrm{d}\xi. \tag{6.27}$$

Definition 6.2. If $f(z)$ is single-valued holomorphic in a simply-connected domain, the integral

$$\int_a^z f(\xi)\,\mathrm{d}\xi \tag{6.28}$$

defined within that domain depends only on the initial point and final point of the path, and does not depend on the path of integration between a and z. In that situation, the function

$$F(z) = \int_a^z f(\xi)\,\mathrm{d}\xi, \tag{6.29}$$

which is uniquely determined, is called the **primitive function** or the **indefinite integral** of $f(z)$.

The primitive function $F(z)$ changes continuously when z is changed. The amount of change may be roughly proportional to the change of the length of the path of integration when the change of z is infinitesimally small. Therefore $F(z)$ is continuous and differentiable, which is to say, it is holomorphic. We will now describe the above in the form of the following theorem.

Theorem 6.5. *When $f(z)$ is single-valued holomorphic in the simply-connected domain D, its indefinite integral $F(z)$ is holomorphic and satisfies*

$$\frac{\mathrm{d}F(z)}{\mathrm{d}z} = f(z). \tag{6.30}$$

Proof. Taking $z + \Delta z$ as a point within D (a regular point of $f(z)$), one can obtain

$$F(z + \Delta z) - F(z) = \int_z^{z+\Delta z} f(\xi)\,\mathrm{d}\xi$$

$$= f(z)\Delta z + \int_z^{z+\Delta z} \{f(\xi) - f(z)\}\,\mathrm{d}\xi. \tag{6.31a}$$

$f(z)$ is continuous, so if an appropriate $\delta(> 0)$ is chosen for any arbitrary positive number ε, it satisfies $|f(\xi) - f(z)| < \varepsilon$ for all values of ξ with

$|\xi - z| < \delta$. Therefore,

$$\left| \frac{F(z + \Delta z) - F(z)}{\Delta z} - f(z) \right| = \left| \frac{1}{\Delta z} \int_z^{z+\Delta z} \{f(\xi) - f(z)\} \, \mathrm{d}\xi \right|$$

$$\leq \frac{\varepsilon}{|\Delta z|} \left| \int_z^{z+\Delta z} \mathrm{d}\xi \right| = \varepsilon. \qquad (6.31\text{b})$$

Taking $\Delta z \to 0$ ($\delta \to 0$), we get $\varepsilon \to 0$ from the continuity of $f(z)$ and, hence,

$$\lim_{\Delta z \to 0} \frac{F(z + \Delta z) - F(z)}{\Delta z} = f(z). \qquad (6.31\text{c})$$

Therefore, $F(z)$ is differentiable and holomorphic and Eq. (6.30) is satisfied.

$\square$

6.4.3 *Multi-valuedness of logarithmic functions and integration of* $1/z$

The function $f(z) = 1/z$ has a pole at $z = 0$. Given a doubly-connected domain around the origin $z = 0$, the function $f(z)$ is single-valued holomorphic except for $z = 0$. Suppose that the path of integration C_0 has the initial point z_0 ($0 \leq \arg z_0 < 2\pi$) and the final point z ($0 \leq \arg z < 2\pi$), and the argument lies between 0 and 2π, in the same way as z_0 and z (Fig. 6.10(a)). As $1/z$ on C_0 locates in a simply connected single-valued holomorphic domain, the indefinite integral

$$F(z) = \int_{z_0}^z \frac{1}{\xi} \, \mathrm{d}\xi \qquad (6.32)$$

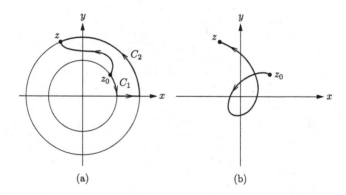

(a)   (b)

Fig. 6.10 Paths of integration for $1/z$.

is uniquely determined. Therefore,

$$F'(z) = \frac{1}{z},$$ (6.33)

so comparing with $(\mathrm{d}/\mathrm{d}z) \log z = 1/z$,

$$F(z) = \log z + a \qquad (a : \text{complex constant}).$$ (6.34)

As shown in Fig. 6.10(a), consider a path of integration (from z_0 to z) that goes back to $|z_0|$ (the point with the argument 0) on a circle of radius $|z_0|$ (path C_1), then moves on the real axis from $|z_0|$ to $|z|$, and finally moves in the positive direction on the circle of radius $|z|$ to the final point z. Integrate along that path. Let the entire path be L_0. Even if the path is changed while the initial point and final point remain unchanged within the single-valued holomorphic domain, the integral value does not change, so the integral along L_0 has the same value as that along C_0, to which is assigned $F(z)$. Then,

$$F(z) = \int_{z_0(L_0)}^{z} \frac{1}{\xi}\mathrm{d}\xi = \int_{C_1} \frac{1}{\xi}\mathrm{d}\xi + \int_{|z_0|}^{|z|} \frac{1}{x}\mathrm{d}x + \int_{C_2} \frac{1}{\xi}\mathrm{d}\xi.$$

The corresponding integrals are

$$\int_{C_1} \frac{1}{\xi}\mathrm{d}\xi = -\mathrm{i}\arg z_0 \qquad (0 \le \arg z_0 < 2\pi),$$

$$\int_{|z_0|}^{|z|} \frac{1}{x}\mathrm{d}x = \ln|z| - \ln|z_0|,$$

$$\int_{C_2} \frac{1}{\xi}\mathrm{d}\xi = \mathrm{i}\arg z \qquad (0 \le \arg z < 2\pi).$$

These can be combined to write as

$$\int_{z_0(L_0)}^{z} \frac{1}{\xi}\mathrm{d}\xi = (\ln|z| + \mathrm{i}\arg z) - (\ln|z_0| + \mathrm{i}\arg z_0) \qquad (0 \le \arg z < 2\pi).$$
(6.35)

This is the primitive function of $1/z$. It is important to note here that the path of integration has been uniquely defined to perform the integration.

Next, we will consider another path that also runs from z_0 to z. This path first rotates in a positive direction around the origin and then heads towards z (Fig. 6.10(b)). Let this path be named L_1. The path L_1 rotates around the origin by 2π, so the argument of the initial point satisfies $0 \le$

$\arg z_0 < 2\pi$, and the argument of the final point satisfies $2\pi \leq \arg z < 4\pi$. This integrates as

$$\int_{z_0(L_1)}^{z} \frac{1}{\xi}\mathrm{d}\xi = \int_{\arg \xi = \arg z_0}^{\arg \xi = 2\pi} \frac{1}{\xi}\mathrm{d}\xi + \int_{|z_0|}^{|z|} \frac{1}{x}\mathrm{d}x + \int_{\arg \xi = 2\pi}^{\arg \xi = \arg z} \frac{1}{\xi}\mathrm{d}\xi$$

$$= \mathrm{i}(2\pi - \arg z_0) + \ln|z| - \ln|z_0| + \mathrm{i}(\arg z - 2\pi)$$

$$= (\ln|z| + \mathrm{i}\arg z) - (\ln|z_0| + \mathrm{i}\arg z_0) \qquad (2\pi \leq \arg z < 4\pi).$$

$$(6.36)$$

Furthermore, let the path of integration that rotates n times around the origin in the positive direction be L_n. The integral on L_n can be calculated in the same way as

$$\int_{z_0(L_n)}^{z} \frac{1}{\xi}\mathrm{d}\xi = (\ln|z| + \mathrm{i}\arg z) - (\ln|z_0| + \mathrm{i}\arg z_0),$$

$$2n\pi \leq \arg z < 2(n+1)\pi. \qquad (6.37)$$

In the above calculation (6.37), the part of $\ln|z| + \mathrm{i}\arg z$ [$2n\pi \leq \arg z < 2(n+1)z$] corresponds to the multi-valuedness of $\log z$. Thus it can be seen that when integrating $1/\xi$ from z_0 to z, the number of times the path rotates around the origin is associated with the multi-valuedness of the logarithmic function $\log z$.

6.5 Residue

6.5.1 *Definition of residue and residue theorem*

Consider a closed Jordan curve C on a complex plane and assume that there is only one pole interior to C. In that case, the integral along a contour path that makes one rotation along C in the positive direction may give a non-zero value.

Example 6.3. Consider $1/z^2$ which has a (second order) pole $z = 0$. Integrate it on the contour path of a circle $|z| = r$. $z = r\,\mathrm{e}^{\mathrm{i}\theta}$ and $\mathrm{d}z = \mathrm{i}\,r\,\mathrm{e}^{\mathrm{i}\theta}\mathrm{d}\theta$, so

$$\oint_{|z|=r} \frac{1}{z^2}\mathrm{d}z = \int_0^{2\pi} \frac{\mathrm{i}\,r\,\mathrm{e}^{\mathrm{i}\theta}}{r^2\,\mathrm{e}^{\mathrm{i}2\theta}}\mathrm{d}\theta = \frac{\mathrm{i}}{r}\int_0^{2\pi} \mathrm{e}^{-\mathrm{i}\theta}\mathrm{d}\theta = -\left.\frac{\mathrm{e}^{-\mathrm{i}\theta}}{r}\right|_0^{2\pi} = 0. \qquad (6.38)$$

In this case, the value of the integral is 0.

Example 6.4. Consider $1/z$ which has a (first-order) pole $z = 0$, and perform the same integration;

$$\oint_{|z|=r} \frac{\mathrm{d}z}{z} = \int_0^{2\pi} \frac{\mathrm{i}\,r\,\mathrm{e}^{\mathrm{i}\theta}}{r\,\mathrm{e}^{\mathrm{i}\theta}}\mathrm{d}\theta = \mathrm{i}\int_0^{2\pi} \mathrm{d}\theta = 2\pi\mathrm{i}. \qquad (6.39)$$

The result of the integration does not depend upon the radius r and is $2\pi i$ instead of 0.

Example 6.5. In the equation

$$f(z) = \frac{z}{(z-1)(z-2)} = -\frac{1}{z-1} + \frac{2}{z-2}, \tag{6.40}$$

$z = 1$ and $z = 2$ are first-order poles. Consider the three closed paths of integration:

C_1: A circle centered at $z = 1$ with a radius 0.5,
C_2: A circle centered at $z = 2$ with a radius 0.5,
C_3: A circle centered at $z = 0$ with a radius 3.

Contour path C_1: There is only single pole $z = 1$ interior to C_1. A function $2/(z-2)$ is holomorphic at any point interior to and on C_1, so the integral is 0. On the path C_1, $z - 1 = 0.5\,e^{i\theta}$, so

$$\oint_{C_1} f(z)\,dz = -\int_0^{2\pi} \frac{i0.5\,e^{i\theta}}{0.5\,e^{i\theta}}\,d\theta + \oint_{C_1} \frac{2}{z-2}\,dz = -i\int_0^{2\pi} d\theta + 0 = -2\pi i.$$

Contour path C_2: A function $1/(z-1)$ is holomorphic at any point interior to and on C_2, so the integral is 0. Taking $z - 2 = 0.5e^{i\theta}$,

$$\oint_{C_2} f(z)\,dz = -\oint_{C_2} \frac{dz}{z-1} + 2\int_0^{2\pi} \frac{i0.5\,e^{i\theta}}{0.5\,e^{i\theta}}\,d\theta = 0 + 2i\int_0^{2\pi} d\theta = 4\pi i.$$

Contour path C_3: From Th. 6.4, adding together the contributions from two poles produces

$$\oint_{C_3} f(z)\,dz = \oint_{C_1} f(z)\,dz + \oint_{C_2} f(z)\,dz = 2\pi i. \tag{6.41}$$

Now let us examine the following examples.

Example 6.6. Integrate $z^{1/2}$ on a circle with radius r where the argument of z changes from 0 to 2π. $z^{1/2}$ is a two-valued function, so the path of integration is not closed on Riemann surface. Then, the argument of $z^{1/2}$ must be determined. Integration is performed on the circle $|z| = r$ from $\arg z = 0$ to $\arg z = 2\pi$. On the circle $z = re^{i\theta}$, so

$$\int_C z^{1/2}dz = ir^{3/2}\int_0^{2\pi} e^{i\theta/2}e^{i\theta}d\theta = ir^{3/2}\frac{1}{3i/2}e^{i3\theta/2}\Big|_{\theta=0}^{2\pi}$$

$$= \frac{2r^{3/2}}{3}(-1-1) = -\frac{4}{3}r^{3/2}. \tag{6.42}$$

Example 6.7. Integrate $f(z) = \exp(1/z)$ on a unit circle $|z| = 1$, rotating around in the positive direction (counterclockwise).

$$\exp\left(\frac{1}{z}\right) = 1 + \frac{1}{1!}\frac{1}{z} + \frac{1}{2!}\frac{1}{z^2} + \cdots + \frac{1}{n!}\frac{1}{z^n} + \cdots . \tag{6.43}$$

Integrating the above for each term produces

$$\oint_{|z|=1} \frac{dz}{z^n} = \begin{cases} 0 & (n \neq 1) \\ 2\pi i & (n = 1) \end{cases},$$

so

$$\oint_{|z|=1} \exp\left(\frac{1}{z}\right) dz = 2\pi i. \tag{6.44}$$

Note that $z = 0$ is an (isolated) essential singularity.

From the above, it can be seen that there is a case in which the integration of one rotation around an (isolated) essential singularity can produce a non-zero finite value. We define residue and related properties as follows.

Definition 6.3 (Residue). Suppose that a closed Jordan curve C has only one isolated singularity inside it, and that, except this singularity, $f(z)$ is holomorphic and interior to and on C. Integrating along a contour path C in the positive direction with $z = z_0$ as the only isolated singularity inside C, the integral

$$\frac{1}{2\pi i} \oint_C f(z)\,dz = A(z_0) \tag{6.45}$$

is called the **residue** of $f(z)$ at $z = z_0$. The residue can be written in forms such as $\operatorname{Res} f(z)|_{z=z_0}$, $\operatorname{Res} f(z_0)$, or $\operatorname{Res}(z_0)$. The residue is determined only by the function $f(z)$ and the singularity z_0.

Theorem 6.6. *At the isolated singularity z_0, if the integral*

$$\lim_{z \to z_0} (z - z_0) f(z) = A \tag{6.46}$$

takes a finite definite value, then A is the residue of $f(z)$ at $z = z_0$.

Proof. Suppose $\lim_{z \to z_0} (z - z_0) f(z) = A$ is a finite definite value. For any positive number ε, if an appropriate positive number δ is chosen such as $|z - z_0| < \delta$, it can be stated that

$$|(z - z_0) f(z) - A| < \varepsilon. \tag{6.47a}$$

Take a contour path C that rotates once around $z = z_0$ and includes z_0 as a single isolated singularity of $f(z)$. The contour path C is transformed to a circle centered at z_0 with an appropriate radius ρ, then

$$z - z_0 = \rho e^{i\theta}. \tag{6.47b}$$

This is a continuous transformation of a closed curve in the holomorphic domain of $f(z)$, so the value of the integral does not change, and

$$\oint_C f(z)\,dz = \oint_{|z-z_0|=\rho} f(z)\,dz. \tag{6.47c}$$

Furthermore, $dz = i\rho e^{i\theta} d\theta = i(z - z_0)\,d\theta$, so

$$\left| \oint_{|z-z_0|=\rho} f(z)\,dz - 2\pi i A \right| = \left| i \int_0^{2\pi} (z - z_0) f(z)\,d\theta - 2\pi i A \right|$$

$$= \left| \int_0^{2\pi} \{(z - z_0)\,f(z) - A\}\,d\theta \right| \le \int_0^{2\pi} |(z - z_0)\,f(z) - A|\,d\theta$$

$$< \varepsilon \int_0^{2\pi} d\theta = 2\pi\varepsilon. \tag{6.47d}$$

Taking the limit $\varepsilon \to 0$, the right-hand side of Eq. (6.47d) becomes 0, so

$$\oint_C f(z)\,dz = 2\pi i A. \tag{6.47e}$$

Therefore, the right-hand side A of Eq. (6.46) is the residue of the complex function $f(z)$ at $z = z_0$. □

However, it is not always possible to find the residue using the method given in Th. 6.6. Let us consider some examples.

Example 6.8. Assuming $c_{-1} \ne 0$ and let $f(z)$ be as follows near $z = z_0$:

$$f(z) = \frac{c_{-1}}{z - z_0} + c_0 + c_1(z - z_0) + c_2(z - z_0)^2 + \cdots. \tag{6.48a}$$

In this case,

$$\lim_{z \to z_0} (z - z_0)\,f(z) = c_{-1}, \tag{6.48b}$$

so the residue of $f(z)$ at $z = z_0$ is c_{-1} (z_0: the first-order pole). This is in the case that Th. 6.6 is applicable.

Example 6.9. Let $c_{-k} \ne 0$, $k \ge 2$, and $f(z)$ be expanded near $z = z_0$ as

$$f(z) = \frac{c_{-k}}{(z - z_0)^k} + \cdots + \frac{c_{-1}}{(z - z_0)} + c_0 + c_1(z - z_0) + \cdots \tag{6.49a}$$

(z_0: the kth order pole). In this case,

$$(z - z_0)\,f(z) \to \infty \qquad (z \to z_0). \tag{6.49b}$$

Thus, the method of Th. 6.6 cannot be applied to find the residue for the kth order pole ($k \ne 1$).

When $z = z_0$ is a removable singularity of $f(z)$,

$$\lim_{z \to z_0} f(z) = f_0 \qquad \text{(finitely determinate)} \qquad (6.50)$$

so it follows that

$$\lim_{z \to z_0} (z - z_0) f(z) = 0. \qquad (6.51)$$

Therefore, applying the same procedure as in the proof of Th. 6.6 for closed contour path C taking one rotation around $z = z_0$, it can be seen that[2]

$$\oint_C f(z) \, \mathrm{d}z = 0. \qquad (6.52)$$

If the isolated singularity is an (isolated) essential singularity, $\lim_{z \to z_0} (z - z_0) f(z)$ is not a finite determinate value.

From the above, $\lim_{z \to z_0} (z - z_0) f(z)$ becomes a finite determinate value under Th. 6.6 only when $z = z_0$ is the first-order pole.

When the isolated singularity is an kth order pole, (where $k \neq 1$), the residue generally is not 0 or infinite. As $\oint_{|z|=1} (1/z^n) \, \mathrm{d}z = 0$ $(n \neq 1)$, by direct integration of $f(z)$ in Example 6.9, we can obtain

$$\frac{1}{2\pi i} \oint_{|z-z_0|=\rho} f(z) \, \mathrm{d}z = c_{-1}. \qquad (6.53)$$

This is the residue. In other words, even if the isolated singularity $z = z_0$ is the kth order pole $(k \neq 1)$, the residue at $z = z_0$ is the coefficient c_{-1} of the -1th order term.

Residue of the order k pole: If $z = a$ is a kth order pole,

$$\operatorname{Res} f(a) = c_{-1} = \lim_{z \to a} \frac{1}{(k-1)!} \frac{\mathrm{d}^{k-1}}{\mathrm{d}z^{k-1}} \{(z-a)^k f(z)\}. \qquad (6.54)$$

This can be shown directly from the concrete form of the Eq. (6.49a).

Example 6.10.

$$f(z) = \frac{z}{(z-1)(z-2)}. \qquad (6.55a)$$

Poles are $z = 1$ and $z = 2$. The corresponding residues are, respectively,

$$\lim_{z \to 1} (z-1) f(z) = -1, \qquad \lim_{z \to 2} (z-2) f(z) = 2. \qquad (6.55b)$$

These results of residues were found in Example 6.5 as complex integrals along the contour paths C_1 and C_2.

---

[2]According to Morera's theorem explained later, this fact is equivalent to the holomorphy of the function $f(z)$. In other words, this guarantees that singularity at such a point can be removable.

Theorem 6.7 (Residue theorem). *If there are N poles z_k ($k = 1, 2, \cdots N$) interior to the closed contour path C, which is traversed in the positive direction, and, except these poles, $f(z)$ is single-valued holomorphic interior to and on the path (Fig. 6.11), then*

$$\oint_C f(z)\,\mathrm{d}z = 2\pi\mathrm{i}\sum_{k=1}^{N}\operatorname{Res}(z_k). \tag{6.56}$$

Proof. We will present only the outline of the proof here. The closed contour path C is divided into contour paths C_k rotating around each pole z_k and paths connecting C_k's. The integral on each C_k is calculated as usual. The path connecting C_k's must pass in both directions. Since the domain of this go–and–return path is single-valued holomorphic, the values of integration in both directions along the path connecting each C_k cancel out, and the net value is 0. As a result, the value of the whole integral becomes the sum of the residues of the poles interior to the closed Jordan curve. $\qquad\square$

6.5.2 *Residue of the point at infinity*

In Sec. 6.5.1, we discussed the residue of an isolated singularity at a finite distance from the origin. In Definition 3.1, the point at infinity $z = \infty$ is introduced and it is handled in the same way as a general point z. Here, we will define the residue of the point at infinity and consider its meaning and nature.

Definition 6.4 (Residue of the point at infinity). For a finite positive number R, let $f(z)$ be holomorphic for $R < |z| < +\infty$, which is to say, there is no singularity, exterior to a circle of radius R except the point at infinity.

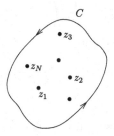

Fig. 6.11   The singular points are distributed.

The Integral

$$-\frac{1}{2\pi i} \oint_{|z|=\rho>R} f(z)\,dz \qquad (6.57)$$

is called the residue of $f(z)$ at $z = \infty$, and expressed as $\operatorname{Res} f(z)\ |_{z=\infty}$, $\operatorname{Res} f(\infty)$, or $\operatorname{Res}(\infty)$, etc. Here, the contour path goes on the circle $|z| = \rho$ in the **positive direction**.

When the rotation on the circle $|z| = \rho$ moves in the positive direction, seen from the origin, the point at infinity will always appear to the right. Therefore, "if the point at infinity is considered as the center, the rotation is in the negative direction". By the above definition, therefore, the coefficient has a negative sign in Eq. (6.57).

Theorem 6.8. *If, when $f(z)$ is holomorphic in the range $R < |z| < \infty$ for a finite positive number R, $\lim_{z\to\infty} zf(z)$ is finitely determinate, then*

$$\operatorname{Res} f(\infty) = -\lim_{z\to\infty} zf(z). \qquad (6.58)$$

Proof. Substituting the variable $z = 1/\zeta$ $\left(dz = \frac{-1}{\zeta^2}d\zeta\right)$, and considering a circle with radius ρ $(> R)$,

$$\operatorname{Res}(\infty) = -\frac{1}{2\pi i} \oint_{|z|=\rho} f(z)\,dz = \frac{1}{2\pi i} \oint_{|\zeta|=1/\rho} f\left(\frac{1}{\zeta}\right) \frac{-1}{\zeta^2}\,d\zeta. \qquad (6.59a)$$

Here, the contour path that goes on the circle $|z| = \rho$ in the positive direction on the z plane is projected onto a contour path that goes on the circle $|\zeta| = 1/\rho$ in the negative direction on the ζ plane. In the final equation, that contour path is rewritten into a contour path that goes on the circle $|\zeta| = 1/\rho$ in the positive direction and multiplied by the sign (-1).

As $f(1/\zeta)$ is holomorphic for $0 < |\zeta| < 1/R$, with the exception of $\zeta = 0$, if there is a singularity in $F(\zeta) \equiv -f(1/\zeta)(1/\zeta^2)$ it is at $\zeta = 0$. Thus,

$$\frac{1}{2\pi i} \oint_{|\zeta|=1/\rho} f\left(\frac{1}{\zeta}\right) \frac{-1}{\zeta^2}\,d\zeta = \frac{1}{2\pi i} \oint_{|\zeta|=1/\rho} F(\zeta)\,d\zeta = \operatorname{Res} F(0) \qquad (6.59b)$$

is reduced to, if $\lim_{\zeta\to 0} \zeta F(\zeta)$ is finite,

$$\operatorname{Res} F(0) = \lim_{\zeta\to 0} \zeta F(\zeta). \qquad (6.59c)$$

Therefore,

$$\operatorname{Res} F(0) = \lim_{\zeta\to 0} \zeta F(\zeta) = -\lim_{z\to\infty} \frac{1}{z} f(z)\, z^2 = -\lim_{z\to\infty} zf(z). \qquad (6.59d)$$

$\square$

Suppose $f(z)$ is expanded as

$$f(z) = \sum_{n=-\infty}^{\infty} c_n z^n \tag{6.60}$$

in the region $R \le |z| \le \infty$. Considering the fact,

$$\oint_{|z|=\rho>0} z^n \, dz = \begin{cases} 0 & (n \ne -1) \\ 2\pi i & (n = -1) \end{cases}, \tag{6.61}$$

it follows that

$$\operatorname{Res} f(\infty) = -c_{-1}. \tag{6.62}$$

From this, it can be understood that even if the point at infinity $z = \infty$ is a holomorphic point for $f(z)$, the residue $\operatorname{Res}(\infty)$ need not necessarily be 0. That fact should be notified.

Theorem 6.9. *A closed Jordan curve C contains all singular points of $f(z)$ at a finite distance inside (Fig. 6.12). Let the singular points of $f(z)$ interior to C be $\{z_k\}$, the residues at each of them are A_k respectively, and the residue at the point at infinity be B_∞. $f(z)$ is holomorphic everywhere except these singular points. In that case,*

$$\sum_k A_k + B_\infty = 0. \tag{6.63}$$

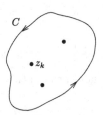

Fig. 6.12   Contour C and singular points.

Proof. The complex integral of the contour C is

$$\oint_C f(z) \, dz = 2\pi i \sum_k A_k. \tag{6.64a}$$

On the other hand, the left-hand side also gives the residue at the point of infinity, according to Definition (6.57).

$$\oint_C f(z) \, dz = -2\pi i B_\infty. \tag{6.64b}$$

Therefore, $\sum A_k + B_\infty = 0$. □

This means that even if $z = \infty$ is a holomorphic point, it generally holds that $B_\infty \neq 0$. Conversely, even if the point at infinity $z = \infty$ is a singular point of $f(z)$, $B_\infty = 0$ if there is no other singular point of $f(z)$.

Example 6.11. For the function

$$f(z) = e^z, \tag{6.65a}$$

$z = \infty$ is an essential singularity and $f(z)$ is holomorphic at all other z points. From Th. 6.9, the residue of $z = \infty$ is 0. Expanding $f(z)$ with the power series of z,

$$e^z = 1 + \frac{1}{1!}z + \frac{1}{2!}z^2 + \cdots + \frac{1}{n!}z^n + \cdots . \tag{6.65b}$$

The term z^{-1} does not appear, so the residue at $z = \infty$ is certainly 0.

6.6 Application of complex integrals

6.6.1 *Application of residue theorem (calculation of definite integrals)*

If a complex integral is used, it may become easier in many cases to execute a definite integral that may be difficult in real analysis.

Example 6.12.

$$I = \int_0^{2\pi} \frac{d\theta}{5 - 4\cos\theta}. \tag{6.66}$$

Taking $z = e^{i\theta}$,

$$\cos\theta = \frac{1}{2}\left(z + \frac{1}{z}\right), \qquad dz = ie^{i\theta}d\theta = iz\,d\theta.$$

When θ varies from 0 to 2π, z goes around the unit circle $|z| = 1$. Therefore, I can be rewritten as

$$I = \oint_{|z|=1} \frac{dz}{iz} \frac{1}{5 - 2\left(z + \frac{1}{z}\right)} = i\oint_{|z|=1} dz \frac{1}{2z^2 - 5z + 2}$$

$$= \frac{i}{2} \oint_{|z|=1} dz \frac{1}{\left(z - \frac{1}{2}\right)(z - 2)}.$$

The only singular point interior to the contour path is the first-order pole $z = 1/2$. So the residue can be evaluated and the resulting integral is

$$I = \frac{i}{2} \cdot 2\pi i \,\text{Res}\left(\frac{1}{2}\right) = -\pi \lim_{z \to 1/2}\left(z - \frac{1}{2}\right) \cdot \frac{1}{\left(z - \frac{1}{2}\right)(z - 2)} = \frac{2\pi}{3}.$$

$$\tag{6.67}$$

Example 6.13.

$$I = \frac{1}{2} \oint_{|z|=2} \frac{|dz|}{|z-1|^2}. \tag{6.68}$$

Taking $z = 2e^{i\theta}$, $dz = 2ie^{i\theta}d\theta$, $|dz| = 2\,d\theta$. It can be rewritten as

$$I = \int_0^{2\pi} d\theta \frac{1}{(2e^{i\theta}-1)(2e^{-i\theta}-1)} = \int_0^{2\pi} \frac{d\theta}{5-4\cos\theta}.$$

This is in the previous Example 6.12, so

$$I = \frac{2\pi}{3}. \tag{6.69}$$

To do more general integrals, we will prove the following lemma.

Lemma 6.1 (Jordan's lemma). *Assume that when $|z| \to \infty$ on the upper half of the z plane ($0 \le \arg z \le \pi$), $f(z)$ uniformly converges to 0. Taking $a > 0$ in this situation,*

$$I_R = \int_{C_R} e^{iaz} f(z)\,dz \to 0 \qquad (R \to \infty). \tag{6.70}$$

Here, the path of integration C_R is a semicircle in which a point $z = Re^{i\theta}$ on a circle of radius R moves from $\theta = 0$ to π on the upper half of the plane (Fig. 6.13).

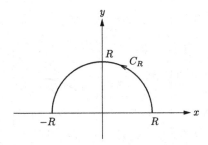

Fig. 6.13   Path of integration C_R (semicircle of radius R) in Jordan's lemma.

Proof. Evaluate $|I_R|$ in Eq. (6.70) as follows:

$$|I_R| = \left| \int_{C_R} e^{iaz} f(z)\,dz \right| \le \int_{C_R} |e^{iaz}||f(z)||dz|. \tag{6.71a}$$

Taking $z = R\mathrm{e}^{\mathrm{i}\theta}$ here, $\mathrm{d}z = \mathrm{i}R\mathrm{e}^{\mathrm{i}\theta}\mathrm{d}\theta$, $|\mathrm{d}z| = R\,\mathrm{d}\theta$, $|\mathrm{e}^{\mathrm{i}az}| = \mathrm{e}^{-aR\sin\theta}$. Substituting them, one can evaluate them as

$$|I_R| \leq \int_0^\pi \mathrm{e}^{-aR\sin\theta}|f(R\mathrm{e}^{\mathrm{i}\theta})|R\,\mathrm{d}\theta. \tag{6.71b}$$

$|f(z)|$ uniformly approaches 0 when $|z| \to \infty$ and, hence, we obtain

$$|f(z)| < \varepsilon \tag{6.71c}$$

on the path, provided a radius R is set large enough with an arbitrary value $\varepsilon > 0$. Therefore,

$$|I_R| \leq \varepsilon R \int_0^\pi \mathrm{e}^{-aR\sin\theta}\mathrm{d}\theta = 2\varepsilon R \int_0^{\pi/2} \mathrm{e}^{-aR\sin\theta}\mathrm{d}\theta. \tag{6.71d}$$

Furthermore, $\sin\theta \geq 2\theta/\pi$ in the range $0 \leq \theta \leq \pi/2$ (Fig. 6.14), so

$$|I_R| \leq 2\varepsilon R \int_0^{\pi/2} \mathrm{e}^{-(2aR/\pi)\theta}\mathrm{d}\theta = \frac{\varepsilon\pi}{a}\left(1 - \mathrm{e}^{-aR}\right) \leq \frac{\pi}{a}\varepsilon. \tag{6.71e}$$

ε can be made smaller to any degree by increasing the radius R, such that

$$\varepsilon \to 0 \qquad (R \to \infty).$$

Then we obtain

$$\lim_{R\to\infty} |I_R| = 0. \tag{6.71f}$$

$\square$

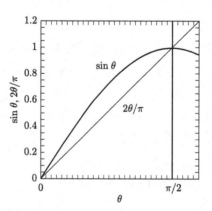

Fig. 6.14   Graph showing $\sin\theta \geq 2\theta/\pi$ to prove Jordan's lemma.

In the above, we considered the upper half of the z plane. If $f(z)$ uniformly approaches 0 as $|z| \to \infty$ on the lower half plane ($2\pi \geq \arg z \geq \pi$), then we set $a > 0$ and consider an exponential function $\mathrm{e}^{-\mathrm{i}az}$. This produces

$$\int_{C'_R} \mathrm{e}^{-\mathrm{i}az} f(z)\,\mathrm{d}z \to 0 \qquad (R \to \infty), \tag{6.72}$$

where the path C'_R is a semicircle in which $z = R\,\mathrm{e}^{\mathrm{i}\theta}$ on a circle of radius R moves from $\theta = \pi$ to 2π in the lower half plane. Then, Jordan's lemma is satisfied in the same way.

When performing a complex integral along the real axis ($\int_{-R}^{R} \mathrm{d}z$), adding a semicircular path of integration such as C_R or C'_R creates a closed contour path, so that Jordan's lemma can be used to lead to the calculation of the residue. It is also possible to do the same to much more different paths of integration, to use the transformation of Jordan's lemma.

The following is an example that brings it into the standard Jordan's lemma with a small transformation.

Example 6.14.

$$I = \int_0^\infty \frac{x \sin ax}{x^2 + 1}\,\mathrm{d}x \qquad (a > 0). \tag{6.73}$$

This can be rewritten using $\sin ax = (\mathrm{e}^{\mathrm{i}ax} - \mathrm{e}^{-\mathrm{i}ax})/(2\mathrm{i})$, as

$$I = \frac{1}{2\mathrm{i}} \int_0^\infty \frac{x}{x^2 + 1}(\mathrm{e}^{\mathrm{i}ax} - \mathrm{e}^{-\mathrm{i}ax})\,\mathrm{d}x = \frac{1}{2\mathrm{i}} \int_{-\infty}^\infty \frac{x\,\mathrm{e}^{\mathrm{i}ax}}{x^2 + 1}\,\mathrm{d}x. \tag{6.74}$$

Let us write a function $f(z)$ as

$$f(z) = \frac{z}{z^2 + 1}. \tag{6.75}$$

The complex integral along the path C_R (Fig. 6.13), according to Jordan's lemma,

$$\int_{C_R} \mathrm{e}^{\mathrm{i}az} f(z)\,\mathrm{d}z \tag{6.76}$$

converges to 0 at the limit $R \to \infty$. Adding this in, we can write

$$I = \lim_{R \to \infty} \frac{1}{2\mathrm{i}} \oint_{C_1} \frac{z\,\mathrm{e}^{\mathrm{i}az}}{z^2 + 1}\,\mathrm{d}z. \tag{6.77}$$

The contour path C_1 moves from $-R$ to R on the real axis, after which it moves on the semicircle C_R to close (Fig. 6.15). In Eq. (6.77), the pole of the integrand is $z = \pm\mathrm{i}$, but inside C_1 there is only $z = \mathrm{i}$. The residue is

$$\lim_{z \to \mathrm{i}}(z - \mathrm{i}) \frac{z\,\mathrm{e}^{\mathrm{i}az}}{z^2 + 1} = \lim_{z \to \mathrm{i}} \frac{z\,\mathrm{e}^{\mathrm{i}az}}{z + \mathrm{i}} = \frac{\mathrm{e}^{-a}}{2}.$$

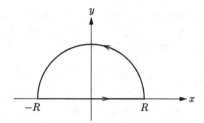

Fig. 6.15   Contour path C_1 in Example 6.14.

Therefore,

$$I = \frac{1}{2i}2\pi i\frac{e^{-a}}{2} = \frac{\pi}{2}e^{-a} \qquad (6.78)$$

is obtained.

Definition 6.5. Assume that a function $f(x)$ is continuous for $a \leq x \leq b$ on the real axis, except a singular point c. If an integral

$$\lim_{\varepsilon \to 0}\left[\int_a^{c-\varepsilon} f(x)\,\mathrm{d}x + \int_{c+\varepsilon}^b f(x)\,\mathrm{d}x\right] \qquad (6.79)$$

converges to a finite value when $\varepsilon > 0$ and a domain with the "same width ε" is excluded on both sides of $z = c$, this is called **Cauchy's principal value (integral)**, or simply, the principal value (integral), and is written as

$$\mathrm{Pv}\int_a^b f(x)\,\mathrm{d}x. \qquad (6.80)$$

Example 6.15.

$$I = \int_0^\infty \frac{\sin x}{x}\,\mathrm{d}x. \qquad (6.81)$$

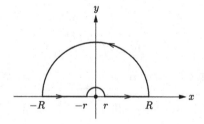

Fig. 6.16   Contour path C_2 in Example 6.15.

This can be rewritten as follows, in the sense of Cauchy's principal value.

$$
I = \lim_{R\to\infty, r\to 0} \int_r^R \frac{\sin x}{x}\,dx = \lim_{R\to\infty, r\to 0} \int_r^R \frac{e^{ix} - e^{-ix}}{2ix}\,dx
$$

$$
= \frac{1}{2i} \lim_{R\to\infty, r\to 0} \left(\int_{-R}^{-r} \frac{e^{ix}}{x}\,dx + \int_r^R \frac{e^{ix}}{x}\,dx \right).
$$

Here, we consider a new integral using the contour path C_2, which has been closed, adding two semicircles with radii R and r on the upper half of the plane (Fig. 6.16).

$$
J = \oint_{C_2} \frac{e^{iz}}{z}\,dz
$$

$$
= \int_{-R}^{-r} \frac{e^{ix}}{x}\,dx + \int_{z=re^{i\theta}, \theta=\pi\to 0} \frac{e^{iz}}{z}\,dz + \int_r^R \frac{e^{ix}}{x}\,dx + \int_{z=Re^{i\theta}, \theta=0\to\pi} \frac{e^{iz}}{z}\,dz
$$

$$
= J_1 + J_2 + J_3 + J_4. \tag{6.82}
$$

The only pole of e^{iz}/z, which is a finite distance from the origin, is $z = 0$, so no poles exist interior to the contour path C_2. Therefore,

$$
J = 0. \tag{6.83}
$$

The sum of the first and third terms of Eq. (6.82) is

$$
J_1 + J_3 = \int_{-R}^{-r} \frac{e^{ix}}{x}\,dx + \int_r^R \frac{e^{ix}}{x}\,dx
$$

so

$$
\lim_{R\to\infty,\ r\to 0} (J_1 + J_3) = 2iI. \tag{6.84}
$$

The integral along the semicircle of radius r can be calculated, in the limit of $r \to 0$, as follows:

$$
\lim_{r\to 0} J_2 = \lim_{r\to 0} \int_{z=re^{i\theta}, \theta=\pi\to 0} \frac{e^{iz}}{z}\,dz = \lim_{r\to 0} i \int_\pi^0 d\theta\, e^{ir\cos\theta - r\sin\theta}
$$

$$
= i \int_\pi^0 d\theta = -i\pi. \tag{6.85}
$$

Note that this is half of the residue of e^{iz}/z at $z = 0$ (multiplied by $-2\pi i$). Readers should consider why that is the case. The integral J_4 on a semicircle of radius R will become 0 in the limit of $R \to \infty$, according to Jordan's lemma.

$$
\lim_{R\to\infty} J_4 = 0. \tag{6.86}
$$

Combining them all,

$$0 = \lim_{R\to\infty, r\to 0} (J_1 + J_3) + \lim_{r\to 0} J_2 + \lim_{R\to\infty} J_4 = 2\mathrm{i}I + (-\mathrm{i}\pi) + 0, \qquad (6.87)$$

so we obtain the result

$$I = \frac{\pi}{2}. \qquad (6.88)$$

In this integral, the semicircle of radius r was closed on the upper half of the plane, but it can also be closed on the lower half. In this case, the pole $z = 0$ of $\mathrm{e}^{\mathrm{i}z}/z$ locates interior to the contour path. If the residue of $z = 0$ and the integral along the semicircle are calculated correctly, they produce the same values.

Example 6.16.

$$I = \int_{-\infty}^{\infty} \frac{\mathrm{d}x}{1 + x^2}. \qquad (6.89)$$

Poles of $1/(1 + z^2)$ are $z = \pm\mathrm{i}$. Consider the contour path C_1 in Example 6.14 (Fig. 6.15). Interior to C_1, there is the first-order pole $z = \mathrm{i}$, so the following integral can be calculated;

$$J = \oint_{C_1} \frac{\mathrm{d}z}{1 + z^2} = 2\pi\mathrm{i}\lim_{z\to\mathrm{i}}(z - \mathrm{i})\frac{1}{1 + z^2} = \pi. \qquad (6.90)$$

Rewriting integral J produces

$$J = \int_{-R}^{R} \frac{\mathrm{d}x}{1 + x^2} + \mathrm{i}\int_{0}^{\pi} \mathrm{d}\theta \frac{R\,\mathrm{e}^{\mathrm{i}\theta}}{1 + R^2\,\mathrm{e}^{2\mathrm{i}\theta}}. \qquad (6.91)$$

The integral of the first term becomes I when $R \to \infty$. The integral of the second term is a quality of order of $1/R$ and becomes 0 when $R \to \infty$. Strictly, this should be demonstrated as

$$\left| \int_{0}^{\pi} \mathrm{d}\theta \frac{R\,\mathrm{e}^{\mathrm{i}\theta}}{1 + R^2\,\mathrm{e}^{2\mathrm{i}\theta}} \right| \leq \int_{0}^{\pi} \mathrm{d}\theta \left| \frac{R\,\mathrm{e}^{\mathrm{i}\theta}}{1 + R^2\,\mathrm{e}^{2\mathrm{i}\theta}} \right| = \int_{0}^{\pi} \mathrm{d}\theta \frac{R}{\sqrt{1 + R^4 + 2R^2\cos 2\theta}}$$

$$\leq \int_{0}^{\pi} \mathrm{d}\theta \frac{R}{R^2 - 1} = \frac{\pi R}{R^2 - 1} \to 0 \qquad (R \to \infty).$$

Combining the above produces

$$I = \pi. \qquad (6.92)$$

Example 6.17.

$$I = \int_{-\infty}^{\infty} \mathrm{e}^{-x^2} \cos 2bx\,\mathrm{d}x. \qquad (6.93)$$

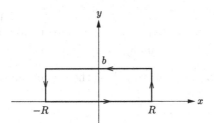

Fig. 6.17 Contour path C_3 in Example 6.17.

Let us integrate e^{-z^2} on a contour path C_3 shown in Fig. 6.17. There are no poles of e^{-z^2} interior to the contour path C_3, so the integral is 0. On the other hand, the integral can be written in detail as

$$
\begin{aligned}
0 &= \oint_{C_3} \mathrm{e}^{-z^2} \mathrm{d}z \\
&= \int_{-R}^{R} \mathrm{e}^{-x^2} \mathrm{d}x + \mathrm{i} \int_{0}^{b} \mathrm{e}^{-(R+\mathrm{i}y)^2} \mathrm{d}y + \int_{R}^{-R} \mathrm{e}^{-(x+\mathrm{i}b)^2} \mathrm{d}x + \mathrm{i} \int_{b}^{0} \mathrm{e}^{-(-R+\mathrm{i}y)^2} \mathrm{d}y \\
&= \int_{-R}^{R} \mathrm{e}^{-x^2} \mathrm{d}x + \int_{R}^{-R} \mathrm{e}^{-x^2+b^2-2\mathrm{i}bx} \mathrm{d}x \\
&\quad + \mathrm{i} \int_{0}^{b} \mathrm{e}^{-R^2+y^2-2\mathrm{i}Ry} \mathrm{d}y + \mathrm{i} \int_{b}^{0} \mathrm{e}^{-R^2+y^2+2\mathrm{i}Ry} \mathrm{d}y.
\end{aligned}
$$

When $R \to \infty$, the integrals of the third and fourth terms of the final expression become 0.[3]

$$
\begin{aligned}
0 &= \int_{-\infty}^{\infty} \mathrm{e}^{-x^2} \mathrm{d}x - \int_{-\infty}^{\infty} \mathrm{e}^{b^2} \mathrm{e}^{-x^2} (\cos 2bx - \mathrm{i}\sin 2bx)\, \mathrm{d}x \\
&= \sqrt{\pi} - \mathrm{e}^{b^2} \int_{-\infty}^{\infty} \mathrm{e}^{-x^2} \cos 2bx\, \mathrm{d}x.
\end{aligned}
$$

Therefore,

$$
I = \mathrm{e}^{-b^2} \sqrt{\pi}. \tag{6.94}
$$

6.6.2 Handling branch points of multi-valued functions in definite integrals

In the integral of a multi-valued function, when the contour path passes near a branch point, we should close the contour path by fully understanding

---

[3]Gauss integral $\int_{-\infty}^{\infty} \mathrm{e}^{-x^2} \mathrm{d}x$ can be calculated as follows. Once the integral is squared, we can calculate it as $\left(\int_{-\infty}^{\infty} \mathrm{e}^{-x^2} \mathrm{d}x \right)^2 = \int_{-\infty}^{\infty} \int_{-\infty}^{\infty} \mathrm{d}x\, \mathrm{d}y\, \mathrm{e}^{-x^2-y^2} = \int_{0}^{\infty} r\, \mathrm{d}r \int_{0}^{2\pi} \mathrm{d}\theta\, \mathrm{e}^{-r^2} = 2\pi \cdot \frac{1}{2} \int_{0}^{\infty} \mathrm{e}^{-t} \mathrm{d}t = \pi$. Then, we obtain $\int_{-\infty}^{\infty} \mathrm{e}^{-x^2} \mathrm{d}x = \sqrt{\pi}$.

Complex Function Theory

the structure of Riemann surface and the argument of the integrand. We will now examine some examples to see how to handle branch points and branch cut in the integral of multi-valued functions.

Example 6.18. Consider integral along the branch cut of an infinite multi-valued function

$$I = \int_0^\infty \frac{x^{p-1}}{1+x} \, dx \qquad (0 < p < 1). \tag{6.95}$$

First, we consider the integral

$$J = \oint_{C_4} \frac{z^{p-1}}{1+z} \, dz \tag{6.96}$$

along the contour path C_4 in Fig. 6.18 (point A is $z = r$, point B is $z = R$, and point C is $z = R\,e^{i\pi}$)).

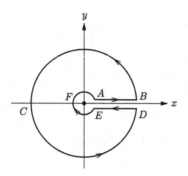

Fig. 6.18   Contour path C_4 in Example 6.18. The branch cut is on the positive side of the real axis, between 0 and ∞.

The arguments of z and z^{p-1} at each point, A through F, on the contour path are as stated in Table 6.1. This contour path does not cross the branch cut, avoiding the branch point $z = 0$, so the path is closed. Since the first-order pole $z = -1 = e^{i\pi}$ lies interior to the contour path C_4 on the same Riemann surface, then we obtain

$$J = 2\pi i \operatorname{Res}(e^{i\pi}) = 2\pi i \, e^{i(p-1)\pi}.$$

Table 6.1   Behavior of the arguments $\arg z$ and $\arg z^{p-1}$ as the point z moves along contour path C_4 (Fig. 6.18).

	A	B	C	D	E	F
$\arg z$	0	0	π	2π	2π	π
$\arg z^{p-1}$	0	0	$\pi(p-1)$	$2\pi(p-1)$	$2\pi(p-1)$	$\pi(p-1)$

Here, it is important to note that the argument of $z = -1$ is π, but not just $-\pi$ or 3π etc. If the argument at point A (then B) is set to be 0, the point C ($z = -1$) is at the angle of π around the origin. Also, J can be rewritten as

$$J = \int_r^R \frac{x^{p-1}}{1+x}dx + \int_0^{2\pi} d\theta\, i\, R\, e^{i\theta} \frac{R^{p-1}e^{i(p-1)\theta}}{1+R\,e^{i\theta}}$$
$$+ \int_R^r \frac{(xe^{2\pi i})^{p-1}}{1+x}dx + \int_{2\pi}^0 d\theta\, i\, r\, e^{i\theta} \frac{r^{p-1}e^{i(p-1)\theta}}{1+r\,e^{i\theta}}.$$

For $D \to E$ (since z goes around the origin by 2π from A ($\arg z = 0$)), the argument is 2π and was written $z = x\,e^{2\pi i}$. In the second term on the right-hand side of the above equation, since $p < 1$, it becomes 0 when $R \to \infty$. The fourth term on the right-hand side, since $p > 0$, becomes 0 when $r \to 0$. Therefore,

$$J \to (1 - e^{i(p-1)2\pi}) \int_0^\infty \frac{x^{p-1}}{1+x}dx \qquad (R \to \infty,\ r \to 0).$$

As a result,

$$\int_0^\infty \frac{x^{p-1}}{1+x}dx = \frac{2\pi i\, e^{i(p-1)\pi}}{1 - e^{i(p-1)2\pi}} = \frac{-\pi}{\sin(p-1)\pi} = \frac{\pi}{\sin p\pi}. \qquad (6.97)$$

Combining these produces

$$I = \frac{\pi}{\sin p\pi}. \qquad (6.98)$$

Let us consider this integral again in Sec. 10.2.2.

Example 6.19. (Schwarz–Christoffel transformation)

Let us consider the following integral along the real axis:

$$z = f(t) = \int_0^t s^{-1/2}(s-1)^{-1/2}ds \qquad (6.99)$$

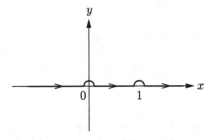

Fig. 6.19   Path of integration by s along the real axis used when considering Schwarz–Christoffel transformation in Example 6.19.

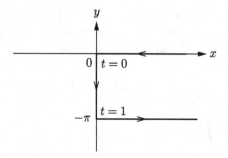

Fig. 6.20   Mapping $z = f(t)$ by Schwarz–Christoffel transformation in Example 6.19.

Let the path of integration along the real axis is traversed in the upper half of the plane, as shown in Fig. 6.19, at the branch points $s = 0$ and $s = 1$.[4] Also, let $\arg s^{1/2} = \arg(s-1)^{1/2} = 0$ at $s > 1$. Then, the arguments of other points on the contour path are also uniquely determined as follows:

$$\arg s^{1/2} = \arg(s-1)^{1/2} = 0 \qquad\qquad (s > 1), \qquad\quad (6.100\text{a})$$

$$\arg s^{1/2} = 0, \quad \arg(s-1)^{1/2} = \pi/2 \qquad (1 > s > 0), \quad (6.100\text{b})$$

$$\arg s^{1/2} = \arg(s-1)^{1/2} = \pi/2 \qquad\qquad (0 > s). \qquad\qquad (6.100\text{c})$$

Therefore, if the arguments are written as above, the integrals can be written as below. $t < 0 : (t = |t|\mathrm{e}^{\mathrm{i}\pi},\ s = r\mathrm{e}^{\mathrm{i}\pi},\ s - 1 = (1 + r)\mathrm{e}^{\mathrm{i}\pi})$

$$z = \int_0^{|t|} (-\mathrm{d}r) r^{-1/2} (1+r)^{-1/2} \mathrm{e}^{-\pi \mathrm{i}} = + \int_0^{|t|} \frac{\mathrm{d}x}{\sqrt{x(1+x)}}. \qquad (6.101\text{a})$$

$0 < t < 1 : (s = r,\ s - 1 = (1 - r)\mathrm{e}^{\mathrm{i}\pi})$

$$z = \int_0^t \mathrm{d}r\, r^{-1/2} (1-r)^{-1/2} \mathrm{e}^{-\mathrm{i}\pi/2} = -\mathrm{i} \int_0^t \frac{\mathrm{d}x}{\sqrt{x(1-x)}}. \qquad (6.101\text{b})$$

$t > 1 : (s = r,\ s - 1 = r - 1)$

$$z = -\mathrm{i} \int_0^1 \frac{\mathrm{d}r}{\sqrt{r(1-r)}} + \int_1^t \frac{\mathrm{d}r}{\sqrt{r(r-1)}} = -\mathrm{i}\pi + \int_1^t \frac{\mathrm{d}x}{\sqrt{x(x-1)}}. \quad (6.101\text{c})$$

From this, $z = f(t)$ is drawn as in Fig. 6.20 on the complex z plane, with t as the parameter.

---

[4]We choose the path moving around the branch points on the upper half-plane. When choosing another path moving around the branch points, the result of the integral would change. Consider what would happen if a point s moves around the branch point on the lower half-plane.

Chapter 7

Cauchy's Integral Formula and Power Series Expansion of Complex Functions

Cauchy's integral formula can be obtained by applying the residue theorem. It is a fruitful formula, and many properties of complex functions can be derived from it. It also extends into the "fundamental theorem of algebra", a variety of basic theorems on the holomorphy of complex functions, and the power series expansion of complex functions.

7.1 Cauchy's integral formula and theorems derived from it

We will now explain Cauchy's integral formula and several important theorems related to it. Morera's theorem explained in this section is the converse of Cauchy's integral theorem. Equivalence of holomorphy of a complex function and Cauchy's integral theorem is derived from this theorem, and the discussion of the holomorphy of complex functions is completed by it.

7.1.1 *Cauchy's integral formula*

Theorem 7.1 (Cauchy's integral formula). *Let D be a domain where $f(z)$ is holomorphic, and C be a contour path that goes around in the positive direction on a closed Jordan curve in D. For any arbitrary point z interior to C (Fig. 7.1), the equation*

$$f(z) = \frac{1}{2\pi i} \oint_C \frac{f(\zeta)}{\zeta - z} d\zeta \qquad (7.1)$$

holds. This is called Cauchy's integral formula.

Proof. Because $f(z)$ is holomorphic in D, the only singular point of $f(\zeta)/(\zeta - z)$ interior to C on the ζ plane is the first-order pole $\zeta = z$.

115

Fig. 7.1   Cauchy's integral formula.

The residue of this function at $\zeta = z$ is

$$\text{Res}\,(z) = \lim_{\zeta \to z}(\zeta - z)\frac{f(\zeta)}{\zeta - z} = f(z), \tag{7.2}$$

so Eq. (7.1) is obtained. □

Example 7.1. (Poisson's integral expression) If $r < R$ for the harmonic function $u(r, \theta)$ ($x = r\cos\theta$, $y = r\sin\theta$), we obtain

$$u(r,\theta) = \frac{1}{2\pi}\int_0^{2\pi} u(R,\phi)\frac{R^2 - r^2}{R^2 - 2Rr\cos(\phi - \theta) + r^2}\,\mathrm{d}\phi. \tag{7.3}$$

This is called *Poisson's integral expression*.

Equation (7.3) can be obtained as follows: In Cauchy's integral formula for a holomorphic function $f(z)$ in a simply-connected domain enclosed by a contour path C

$$f(z) = \frac{1}{2\pi\mathrm{i}}\oint_C \frac{f(\zeta)}{\zeta - z}\,\mathrm{d}\zeta, \tag{7.4}$$

let the contour path C be on a circle of radius R centered at the origin and go around in the positive direction. Letting $z = re^{\mathrm{i}\theta}$, $\zeta = Re^{\mathrm{i}\phi}$ ($r < R$), we

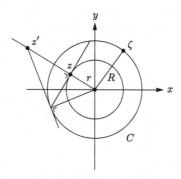

Fig. 7.2   Contour path for Poisson's integral expression (Example 7.1).

obtain

$$f(r\,\mathrm{e}^{\mathrm{i}\theta}) = \frac{1}{2\pi} \int_0^{2\pi} \mathrm{d}\phi \frac{f(R\,\mathrm{e}^{\mathrm{i}\phi})R\,\mathrm{e}^{\mathrm{i}\phi}}{R\,\mathrm{e}^{\mathrm{i}\phi} - r\,\mathrm{e}^{\mathrm{i}\theta}}. \tag{7.5}$$

The mirror image of z with respect to the circle C is $z' = (R^2/r)\mathrm{e}^{\mathrm{i}\theta}$ (Fig. 7.2), which is outside the contour path C. Then,

$$
\begin{aligned}
0 = \frac{1}{2\pi\mathrm{i}} \oint_C \frac{f(\zeta)}{\zeta - z'} \mathrm{d}\zeta &= \frac{1}{2\pi} \int_0^{2\pi} \mathrm{d}\phi \frac{f(R\,\mathrm{e}^{\mathrm{i}\phi})R\,\mathrm{e}^{\mathrm{i}\phi}}{R\,\mathrm{e}^{\mathrm{i}\phi} - (R^2/r)\mathrm{e}^{\mathrm{i}\theta}} \\
&= \frac{1}{2\pi} \int_0^{2\pi} \mathrm{d}\phi \frac{f(R\,\mathrm{e}^{\mathrm{i}\phi})r\,\mathrm{e}^{\mathrm{i}\phi}}{r\,\mathrm{e}^{\mathrm{i}\phi} - R\,\mathrm{e}^{\mathrm{i}\theta}}.
\end{aligned}
\tag{7.6}
$$

Subtracting both sides of Eq. (7.6) from those of Eq. (7.5), we get

$$
\begin{aligned}
f(r\mathrm{e}^{\mathrm{i}\theta}) &= \frac{1}{2\pi} \int_0^{2\pi} \mathrm{d}\phi\, f(R\,\mathrm{e}^{\mathrm{i}\phi}) \left(\frac{R\,\mathrm{e}^{\mathrm{i}\phi}}{R\,\mathrm{e}^{\mathrm{i}\phi} - r\,\mathrm{e}^{\mathrm{i}\theta}} - \frac{r\,\mathrm{e}^{\mathrm{i}\phi}}{r\,\mathrm{e}^{\mathrm{i}\phi} - R\,\mathrm{e}^{\mathrm{i}\theta}} \right) \\
&= \frac{1}{2\pi} \int_0^{2\pi} \mathrm{d}\phi\, f(R\,\mathrm{e}^{\mathrm{i}\phi}) \frac{R^2 - r^2}{R^2 - 2Rr\cos(\phi - \theta) + r^2}.
\end{aligned}
\tag{7.7}
$$

A harmonic function can be a real or imaginary part of some holomorphic function. Then the real part of $f(z) = u(z) + \mathrm{i}v(z)$ can give Eq. (7.3).

Similarly, if both sides of Eq. (7.5) and those of Eq. (7.6) respectively are added, they produce

$$
\begin{aligned}
f(r\mathrm{e}^{\mathrm{i}\theta}) &= \frac{1}{2\pi} \int_0^{2\pi} \mathrm{d}\phi\, f(R\,\mathrm{e}^{\mathrm{i}\phi}) \left(\frac{R\,\mathrm{e}^{\mathrm{i}\phi}}{R\,\mathrm{e}^{\mathrm{i}\phi} - r\,\mathrm{e}^{\mathrm{i}\theta}} + \frac{r\,\mathrm{e}^{\mathrm{i}\phi}}{r\,\mathrm{e}^{\mathrm{i}\phi} - R\,\mathrm{e}^{\mathrm{i}\theta}} \right) \\
&= \frac{1}{2\pi} \int_0^{2\pi} \mathrm{d}\phi\, f(R\,\mathrm{e}^{\mathrm{i}\phi}) \frac{R^2 + r^2 - 2Rr\,\mathrm{e}^{\mathrm{i}(\phi-\theta)}}{R^2 - 2Rr\cos(\phi - \theta) + r^2}.
\end{aligned}
\tag{7.8}
$$

Taking the imaginary part of $f(z) = u(z) + \mathrm{i}v(z)$, we obtain

$$v(r, \theta) = v_0 + \frac{1}{2\pi} \int_0^{2\pi} u(R, \phi) \frac{2Rr\sin(\theta - \phi)}{R^2 - 2Rr\cos(\phi - \theta) + r^2} \mathrm{d}\phi, \tag{7.9}$$

where[1]

$$v_0 = \frac{1}{2\pi} \int_0^{2\pi} v(R, \phi)\, \mathrm{d}\phi = v(0). \tag{7.10}$$

The integral formula (7.1) is the very important theorem, from which a variety of results are derived. This theorem tells that if the value of $f(z)$ on a closed contour path C is given, the value at any arbitrary point interior to C is determined. We will now show several consequences.

---

[1]From Cauchy's integral formula,

$$f(a) = \frac{1}{2\pi\mathrm{i}} \int_{|z-a|=R} \mathrm{d}z \frac{f(z)}{z - a} = \frac{1}{2\pi} \int_0^{2\pi} \mathrm{d}\phi\, f(a + R\,\mathrm{e}^{\mathrm{i}\phi})$$

is obtained. Equation (7.10) is found by letting $a = 0$ here and taking the imaginary parts of both sides.

7.1.2 Maximum modulus principle and Liouville's theorem

Corollary 7.1. *The real part and imaginary part of a holomorphic function do not have the maximum value and minimum value inside the holomorphic domain of this function.*

Proof. Suppose that the real part of $f(z)$ takes the maximum value at one point $z = a$ inside the domain. We choose a contour path C on a circle $|z - a| = r$ in the holomorphic domain and the path C goes around in the positive direction. Assuming the maximum value at $z = a$, then for all θ and enough small r, we get,

$$\operatorname{Re} f(a) > \operatorname{Re} f(a + r\,e^{i\theta}). \tag{7.11a}$$

On the other hand, on C, $z - a = r\,e^{i\theta}$, $dz = i r\,e^{i\theta} d\theta = i(z - a)\,d\theta$, so, rewriting Cauchy's integral formula produces

$$f(a) = \frac{1}{2\pi} \int_0^{2\pi} d\theta\, f(a + re^{i\theta}). \tag{7.11b}$$

This equality holds for the real parts on each side, and for the imaginary parts on each side. The equation for the real part contradicts Eq. (7.11a). Therefore, the real part of $f(z)$ cannot have a maximum value in the domain. The same discussion can be made for the minimum value. Furthermore, the same can be stated for the imaginary part of $f(z)$. $\qquad\square$

Corollary 7.2. *The absolute value of a holomorphic function does not take the maximum value inside the holomorphic domain, and does not take the minimum value other than 0.*

Proof. Let us assume that the absolute value of the holomorphic function $f(z)$ takes the maximum value at a point $z = a$ inside the domain. Then for an adequately small r,

$$|f(a)| > \frac{1}{2\pi} \int_0^{2\pi} d\theta |f(a + re^{i\theta})|. \tag{7.12a}$$

This contradicts the inequality

$$|f(a)| = \frac{1}{2\pi} \left| \int_0^{2\pi} d\theta\, f(a + re^{i\theta}) \right| \leq \frac{1}{2\pi} \int_0^{2\pi} d\theta |f(a + re^{i\theta})|. \tag{7.12b}$$

On the other hand, if $|f(z)|$ takes a non-zero minimum value, then $1/f(z)$ in the neighborhood of $z = a$ is holomorphic and $|1/f(z)|$ has a non-zero maximum value. This also contradicts the assumption. $\qquad\square$

The following few results can already be obtained immediately from Corollary 7.1 and Corollary 7.2.

Corollary 7.3 (Maximum modulus principle). *When $f(z)$ is holomorphic in a closed domain, $|f(z)|$ only takes the maximum value on the boundary of that domain.*

Theorem 7.2 (Liouville's theorem). *If $f(z)$ is holomorphic in the entire complex z plane, "including the point at infinity" (that is to say, it is an entire function), $f(z)$ is a constant.*

Proof. (First proof of Liouville's theorem) As this is a holomorphic function it can be written, for example, as

$$f(z) = c_0 + c_1 z + c_2 z^2 + \cdots . \tag{7.13a}$$

(In fact, a holomorphic function can always be expanded into this kind of power series around a holomorphic point, as shown in Sec. 7.3.1.) Due to holomorphy, a constant M exists, and for all z, $|f(z)| \leq M$. Therefore, for any arbitrary z ($|z| < r$) within a radius r centered at the origin, according to Cauchy's inequality (7.22), which will be described later,

$$|f^{(n)}(z_0)| \leq \frac{n! M}{r^n}. \tag{7.13b}$$

Since $f^{(n)}(0) = n! c_n$ for Eq. (7.13a), then

$$|c_n| \leq \frac{M}{r^n}. \tag{7.13c}$$

Letting $r \to \infty$, we obtain

$$c_n = 0 \qquad (n \neq 0). \tag{7.13d}$$

Thus, all coefficients except c_0 are 0. $\qquad\square$

The proof above used the fact that a holomorphic function can be written as a power series near a holomorphic point, and used Cauchy's inequality. However, we have not yet demonstrated these facts, so we will prove Liouville's theorem in a slightly different way.

Proof. (One further proof of Liouville's theorem) Consider a contour path C that goes in the positive direction along a closed Jordan curve. Furthermore, assume that the path C encircles any two arbitrary points z_1, z_2 (Fig. 7.3). From Cauchy's integral formula,

$$f(z_j) = \frac{1}{2\pi i} \oint_C \frac{f(\zeta)}{\zeta - z_j} d\zeta \qquad (j = 1, 2). \tag{7.14a}$$

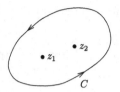

Fig. 7.3   Liouville's theorem.

From that, it follows that

$$f(z_1) - f(z_2) = \frac{1}{2\pi i} \oint_C f(\zeta) \left(\frac{1}{\zeta - z_1} - \frac{1}{\zeta - z_2} \right) d\zeta$$

$$= \frac{z_1 - z_2}{2\pi i} \oint_C \frac{f(\zeta)}{(\zeta - z_1)(\zeta - z_2)} d\zeta. \qquad (7.14\text{b})$$

If C is a circle of adequately large radius R ($\zeta = R\,e^{i\theta}$, $d\zeta = iR\,e^{i\theta}d\theta$),

$$|f(z_1) - f(z_2)| = \frac{|z_1 - z_2|}{2\pi} \left| \int_0^{2\pi} d\theta\, i R\,e^{i\theta} \frac{f(R\,e^{i\theta})}{(R\,e^{i\theta} - z_1)(R\,e^{i\theta} - z_2)} \right|$$

$$\leq \frac{|z_1 - z_2|}{2\pi} \frac{M}{R} \int_0^{2\pi} \frac{d\theta}{|(e^{i\theta} - z_1/R)(e^{i\theta} - z_2/R)|}. \qquad (7.14\text{c})$$

Here, for all values of z, M is a constant satisfying $|f(z)| < M$. Letting $R \to \infty$ here,

$$|f(z_1) - f(z_2)| \to 0. \qquad (7.14\text{d})$$

Since the limit process $R \to \infty$ does not depend on z_1, z_2,

$$f(z_1) = f(z_2) = \text{constant}. \qquad (7.14\text{e})$$

$\square$

7.1.3   *Fundamental theorem of algebra*

Fundamental theorem of algebra (the theorem concerning the roots of algebraic equations) is derived from Liouville's theorem.

Theorem 7.3 (Fundamental theorem of algebra).

$$f(z) = a_0 z^n + a_1 z^{n-1} + \cdots + a_{n-1} z + a_n = 0 \qquad (n \geq 1,\ a_0 \neq 0) \quad (7.15)$$

has n roots (zero points) of complex numbers.

Proof. Suppose $f(z)$ has not even one root. In that case,

$$g(z) = \frac{1}{f(z)} \tag{7.16}$$

is holomorphic for the entire range of the complex plane, including the point at infinity. Therefore, according to Liouville's theorem, $g(z)$ becomes a constant. This conclusion contradicts the assumption, so the hypothesis is not true, and $f(z)$ has at least one root. Letting this root be z_1,

$$f(z) = (z - z_1)\, h(z) \tag{7.17}$$

can be written, and $h(z)$ is an $(n-1)$th order polynomial. If z_1 is known, it is easy to write $h(z)$. Next, we can make the same kind of discussion for the $(n-1)$th order equation $h(z)$. One can continue this below as well. $\square$

7.2 Cauchy's integral theorem and holomorphy

7.2.1 *Goursat's theorem and Morera's theorem*

Further generalizing Cauchy's integral formula, the following formula also holds for the nth derivative of $f(z)$.

Theorem 7.4 (Goursat's theorem). *When $f(z)$ is holomorphic in the simply-connected domain D, take as the contour path a closed Jordan curve C that goes around in the positive direction in D. Choosing an arbitrary point z interior to the contour path C, we have*

$$f^{(n)}(z) = \frac{n!}{2\pi i} \oint_C \frac{f(\zeta)}{(\zeta - z)^{n+1}} d\zeta. \tag{7.18}$$

Proof. From Cauchy's integral formula,

$$
\begin{aligned}
f'(z) &= \lim_{\Delta z \to 0} \frac{f(z + \Delta z) - f(z)}{\Delta z} \\
&= \lim_{\Delta z \to 0} \frac{1}{2\pi i \Delta z} \oint_C \left(\frac{f(\zeta)}{\zeta - z - \Delta z} - \frac{f(\zeta)}{\zeta - z} \right) d\zeta \\
&= \frac{1}{2\pi i} \oint_C \frac{f(\zeta)}{(\zeta - z)^2} d\zeta.
\end{aligned} \tag{7.19}
$$

Therefore, Eq. (7.18) is true when $n = 1$. Employ mathematical induction for $n \geq 2$. Supposing the equation above is true for a certain n, we find

$$
\begin{aligned}
f^{(n+1)}(z) &= \lim_{\Delta z \to 0} \frac{f^{(n)}(z + \Delta z) - f^{(n)}(z)}{\Delta z} \\
&= \frac{n!}{2\pi i} \lim_{\Delta z \to 0} \oint_C \frac{1}{\Delta z} \frac{(\zeta - z)^{n+1} - (\zeta - z - \Delta z)^{n+1}}{(\zeta - z - \Delta z)^{n+1}(\zeta - z)^{n+1}} f(\zeta) \, d\zeta \\
&= \frac{n!}{2\pi i} \oint_C \frac{(n+1)(\zeta - z)^n}{(\zeta - z)^{2(n+1)}} f(\zeta) \, d\zeta \\
&= \frac{(n+1)!}{2\pi i} \oint_C \frac{f(\zeta)}{(\zeta - z)^{n+2}} \, d\zeta.
\end{aligned}
\tag{7.20}
$$

Thus, Eq. (7.18) is true even for $n + 1$. Therefore, Eq. (7.18) is true for all n. $\qquad\square$

If $f(z)$ is holomorphic, it is differentiable any number of times, and its specific form is given by Goursat's theorem. Let us express this in the form of a corollary.

Corollary 7.4. *If $f(z)$ is holomorphic in the domain D, it is differentiable any number of times at every point in D.*

Consider the circle $|\zeta - z| \leq r$ in the holomorphic domain D of $f(\zeta)$. Letting $|f(\zeta)| \leq M$ at every point ζ on the circle $|\zeta - z| = r$, then Goursat's theorem produces

$$
\begin{aligned}
|f^{(n)}(z)| &= \left| \frac{n!}{2\pi i} \oint_{|\zeta - z| = r} \frac{f(\zeta)}{(\zeta - z)^{n+1}} d\zeta \right| \\
&\leq \frac{n!}{2\pi} \oint_{|\zeta - z| = r} |d\zeta| \frac{|f(\zeta)|}{|\zeta - z|^{n+1}} \leq \frac{n!}{2\pi} M \int_0^{2\pi} d\theta \, r \frac{1}{r^{n+1}} \\
&= \frac{n! M}{r^n}.
\end{aligned}
\tag{7.21}
$$

Therefore, the following conclusions can be obtained.

Corollary 7.5 (Cauchy's inequality). *If $f(z)$ is holomorphic in $|z - z_0| \leq r$, and $|f(z)| \leq M$ on $|z - z_0| = r$, then we obtain*

$$
|f^{(n)}(z_0)| \leq \frac{n! M}{r^n}.
\tag{7.22}
$$

We have obtained Goursat's theorem, so we can derive the converse of Cauchy's theorem, which is Morera's theorem.

Theorem 7.5 (Morera's theorem). *Assume that $f(z)$ is continuous in a simply-connected closed domain D. Taking as the contour path any arbitrary closed Jordan curve C in D (Fig. 7.4), then if*

$$\oint_C f(z)\,\mathrm{d}z = 0, \qquad (7.23)$$

$f(z)$ *is holomorphic in D.*

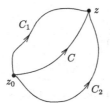

Fig. 7.4   Morera's theorem.

Proof. As was demonstrated with Cauchy's theorem, if the integral is 0 within any arbitrary closed curve, $F(z) = \int_{z_0}^{z} f(\zeta)\,\mathrm{d}\zeta$ is determined, regardless of the path of integration (Fig. 7.4). Therefore, $F(z)$ is a function that is uniquely determined for a given z and is differentiable. So

$$F'(z) = f(z). \qquad (7.24)$$

As $F(z)$ is differentiable, which is to say, it is holomorphic, it is infinitely differentiable according to Goursat's theorem. Therefore $f(z) = F'(z)$ is also differentiable, and holomorphic. □

From the above, we have demonstrated the holomorphy of $f(z)$ and its equivalence with $\oint_C f(z)\,\mathrm{d}z = 0$.

7.2.2   Application of Goursat's theorem

We will now present several definite integrals, using Goursat's theorem.

Example 7.2. Consider the integral

$$I = \int_{-\infty}^{\infty} \frac{\mathrm{d}x}{(1+x^2)^{n+1}} \qquad (n > -1). \qquad (7.25)$$

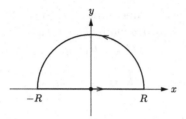

Fig. 7.5   The contour path C_R for Example 7.2.

Choose the contour path C_R (Fig. 7.5) that moves from $-R$ to R, then is closed on the upper half–lane.

$$\oint_{C_R} \frac{\mathrm{d}z}{(1+z^2)^{n+1}} = \oint_{C_R} \frac{\mathrm{d}z}{(z-\mathrm{i})^{n+1}(z+\mathrm{i})^{n+1}}.$$

Applying Goursat's theorem to $f(z) = 1/(z+\mathrm{i})^{n+1}$,

$$\oint_{C_R} \frac{\mathrm{d}z}{(1+z^2)^{n+1}} = \frac{2\pi\mathrm{i}}{n!} f^{(n)}(z=\mathrm{i}) = \frac{\pi(2n)!}{(2^n n!)^2}.$$

On the other hand, when $R \to \infty$, the integral on the semicircle on the upper half plane becomes 0;

$$\lim_{R\to\infty} \oint_{C_R} \frac{\mathrm{d}z}{(1+z^2)^{n+1}} = \int_{-\infty}^{\infty} \frac{\mathrm{d}x}{(1+x^2)^{n+1}}.$$

Therefore, we obtain[2]

$$\int_{-\infty}^{\infty} \frac{\mathrm{d}x}{(1+x^2)^{n+1}} = \pi \frac{1\cdot 3\cdot 5\cdots(2n-1)}{2\cdot 4\cdot 6\cdots(2n)} = \pi\frac{(2n-1)!!}{(2n)!!}. \qquad (7.26)$$

Example 7.3. Calculate the integral

$$I = \int_0^{2\pi} \cos^{2n}\theta\, \mathrm{d}\theta. \qquad (7.27)$$

Transforming this produces

$$I = \int_0^{2\pi} \left(\frac{e^{\mathrm{i}\theta}+e^{-\mathrm{i}\theta}}{2}\right)^{2n}\mathrm{d}\theta = -\frac{\mathrm{i}}{2^{2n}}\oint_{|z|=1}\frac{(1+z^2)^{2n}}{z^{2n+1}}\mathrm{d}z.$$

Here, we used $e^{\mathrm{i}\theta} = z$, $\mathrm{d}\theta = -\mathrm{i}\,\mathrm{d}z/z$. Applying Goursat's theorem to $f(z) = (1+z^2)^{2n}$, we obtain

$$I = -\frac{\mathrm{i}}{2^{2n}}\frac{2\pi\mathrm{i}}{(2n)!}f^{(2n)}(0) = -\frac{\mathrm{i}}{2^{2n}}\cdot 2\pi\mathrm{i}\frac{(2n)!}{(n!)^2} = 2\pi\frac{(2n-1)!!}{(2n)!!}. \qquad (7.28)$$

---

[2]The symbol !! represents the following:

$$(2n)!! = 2n(2n-2)(2n-4)\cdots 4\cdot 2,$$
$$(2n-1)!! = (2n-1)(2n-3)\cdots 3\cdot 1,$$
$$0!! = (-1)!! = 1.$$

7.3   Taylor expansion and Laurent expansion

Until now, we have often used a formula of series expansions for complex functions. This section shows that series expansion is very generally possible for a wide range of functions. Goursat's theorem is used to show this issue. First of all, we will consider the domain of defining the function $f(z)$, and points around which we are trying to expand $f(z)$ into the series.

7.3.1   Taylor expansion (power series expansion around a holomorphic point)

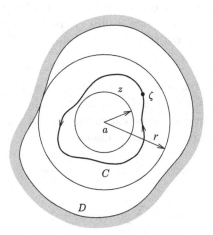

Fig. 7.6   Taylor expansion.

Let $f(z)$ be holomorphic in domain D, and consider two points, z and a in D. Let C be a closed Jordan curve in D, in which z and a lie. Let the point ζ be on C, and satisfy $|\zeta - a| > |z - a|$ (Fig. 7.6). $1/(\zeta - z)$ is expanded into power series as

$$\frac{1}{\zeta - z} = \frac{1}{(\zeta - a) - (z - a)} = \frac{1}{\zeta - a} \sum_{n=0}^{\infty} \left(\frac{z - a}{\zeta - a} \right)^n, \qquad (7.29)$$

which absolutely converges on C because $|z - a| < |\zeta - a|$. Substituting the

Complex Function Theory

above equation into Cauchy's integral formula, we obtain

$$f(z) = \frac{1}{2\pi i} \oint_C \frac{f(\zeta)}{\zeta - z} d\zeta = \frac{1}{2\pi i} \oint_C \sum_{n=0}^{\infty} \frac{(z - a)^n}{(\zeta - a)^{n+1}} f(\zeta) \, d\zeta$$

$$= \frac{1}{2\pi i} \sum_{n=0}^{\infty} (z - a)^n \oint_C \frac{f(\zeta)}{(\zeta - a)^{n+1}} d\zeta. \tag{7.30}$$

According to Goursat's theorem, this is

$$f(z) = \sum_{n=0}^{\infty} \frac{f^{(n)}(a)}{n!} (z - a)^n. \tag{7.31}$$

Note that Eqs. (7.30) and (7.31) hold for all z within the circle (of radius r) that first touches the boundary of the domain D, regardless of the contour path C. This is called the **Taylor expansion (Taylor series)**. The radius r of the circle is the convergence radius of the Taylor expansion. In this way, we have demonstrated that a complex function can be expanded into a Taylor series around its holomorphic point.

Example 7.4. The trigonometric functions $\sin z$ and $\cos z$ are holomorphic in the entire domain within a finite distance from the origin on Gauss plane. Therefore, these functions can be expanded into a Taylor series around the origin, and its convergence radius is infinite:

$$\sin z = \sum_{m=0}^{\infty} (-1)^m \frac{z^{2m+1}}{(2m + 1)!}, \tag{7.32a}$$

$$\cos z = \sum_{m=0}^{\infty} (-1)^m \frac{z^{2m}}{(2m)!}. \tag{7.32b}$$

Example 7.5. The logarithmic function $\log(1 + z)$ can be expanded into a Taylor series as

$$\log(1 + z) = \sum_{n=1}^{\infty} (-1)^{n-1} \frac{z^n}{n}. \tag{7.33}$$

The convergence radius is 1.

Example 7.6. A function can be expanded into a Taylor series around a removable singularity. For example, $\sin z / z$ can be expressed as

$$\frac{\sin z}{z} = \sum_{m=0}^{\infty} (-1)^m \frac{z^{2m}}{(2m + 1)!}. \tag{7.34}$$

The right-hand side converges in the entire z range, including $z = 0$.

Theorem 7.6 (Schwarz's theorem). *When $f(z)$ is holomorphic in $|z| < 1$ and satisfies $|f(z)| < 1$ and $f(0) = 0$,*

$$|f(z)| \leq |z| \tag{7.35}$$

is true at any point z of $|z| < 1$.

Furthermore, if $|f(z_0)| = |z_0|$ at a certain point $z_0 \neq 0$ satisfying $|z_0| < 1$, we can get

$$f(z) = az \tag{7.36}$$

with a complex number a of $|a| = 1$.

Proof. Since $f(0) = 0$, $f(z)$ can be expanded into a Taylor series

$$f(z) = \sum_{n=1}^{\infty} c_n z^n.$$

Defining

$$g(z) = \sum_{n=1}^{\infty} c_n z^{n-1},$$

then $f(z) = zg(z)$ ($|z| < 1$). Since

$$|f(z)| = |z||g(z)| < 1, \tag{7.37}$$

we can get, with a certain positive number r, at $|z| = r < 1$,

$$|g(z)| < \frac{1}{r}. \tag{7.38}$$

From this discussion and the maximum modulus principle, we can see at arbitrary z of $|z| < r$

$$|f(z)| < \frac{1}{r}. \tag{7.39}$$

Since r can be as close to 1 as possible, we get $|g(z)| \leq 1$ ($|z| < 1$). Therefore, from Eq. (7.37)

$$|f(z)| \leq |z| \tag{7.40}$$

is true. (End of the first half of the theorem.)

When $|f(z_0)| = |z_0|$ ($z_0 \neq 0$), $|g(z_0)| = 1$ from Eq. (7.37). Because, $|g(z)| \leq 1$ ($|z| < 1$) from the first half of the theorem, $g(z)$ should be a constant satisfying $|g(z)| = 1$. Then $g(z) = a$ with $|a| = 1$. This means $f(z) = az$. (End of the last half of the theorem.) $\qquad\square$

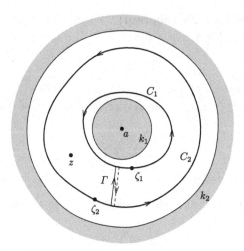

Fig. 7.7   Laurent expansion.

7.3.2   *Laurent expansion*

Let the *annulus* (a domain like ring) D between circles k_1 and k_2, which are both centered at the point a, be a single-value holomorphic domain for $f(z)$. In this case, consider the two contour paths C_1, C_2 in the positive direction, as shown in Fig. 7.7. For all points ζ_1 on C_1 and all points ζ_2 on C_2, let

$$|\zeta_2 - a| > |z - a| > |\zeta_1 - a|. \tag{7.41}$$

Let us consider the $-C_1$ (C_1 in the reverse direction) and C_2 joined by a path Γ which goes back and forth between them. The integrals of Γ in each direction cancel each other out, because $f(z)$ is single-value holomorphic, and

$$f(z) = \frac{1}{2\pi i} \oint_{C_2} \frac{f(\zeta_2)}{\zeta_2 - z} d\zeta_2 + \frac{1}{2\pi i} \oint_{-C_1} \frac{f(\zeta_1)}{\zeta_1 - z} d\zeta_1. \tag{7.42}$$

Furthermore, two series

$$\frac{1}{\zeta_2 - z} = \frac{1}{(\zeta_2 - a) - (z - a)} = \frac{1}{\zeta_2 - a} \sum_{n=0}^{\infty} \left(\frac{z - a}{\zeta_2 - a} \right)^n \tag{7.43a}$$

and

$$\frac{1}{\zeta_1 - z} = \frac{1}{(\zeta_1 - a) - (z - a)} = -\frac{1}{z - a} \sum_{n=0}^{\infty} \left(\frac{\zeta_1 - a}{z - a} \right)^n \tag{7.43b}$$

both absolutely converge. Therefore,

$$f(z) = \sum_{n=0}^{\infty} (z-a)^n \frac{1}{2\pi i} \oint_{C_2} \frac{f(\zeta)}{(\zeta - a)^{n+1}} d\zeta$$

$$+ \sum_{n=0}^{\infty} (z-a)^{-n-1} \frac{1}{2\pi i} \oint_{C_1} (\zeta - a)^n f(\zeta) d\zeta. \qquad (7.44)$$

As $f(z)$ is holomorphic in D, we can modify C_1 and C_2 to make a single contour path C, and obtain

$$f(z) = \sum_{n=0}^{\infty} (z-a)^n \frac{1}{2\pi i} \oint_C \frac{f(\zeta)}{(\zeta - a)^{n+1}} d\zeta \; .$$

$$+ \sum_{n=0}^{\infty} (z-a)^{-n-1} \frac{1}{2\pi i} \oint_C (\zeta - a)^n f(\zeta) d\zeta$$

$$= \sum_{n=-\infty}^{\infty} c_n (z-a)^n. \qquad (7.45)$$

This absolutely converges. Without distinguishing positive or negative n, c_n can be written as

$$c_n = \frac{1}{2\pi i} \oint_C \frac{f(\zeta)}{(\zeta - a)^{n+1}} d\zeta \qquad (-\infty < n < \infty). \qquad (7.46)$$

Equation (7.45) is called the **Laurent expansion (Laurent series)**. Note that the Laurent expansion is often done more directly and easily, without using Eq. (7.46). Look at Examples 7.7 and 7.8 for specific cases.

The part with $n < 0$, which is of the negative power, is called the **principal part** of the Laurent expansion. When the Laurent expansion (7.45) is valid in the neighborhood of $z = a$, except $z = a$ itself, and the highest power of the principal part is $(z-a)^{-n}$, $z = a$ is an nth order pole. When the Laurent expansion (7.45) is possible in the neighborhood of $z = a$ except $z = a$, and the principal part continues infinitely (*i.e.* c_n are, at least, not always 0 when $n \to \infty$), the point a is an (isolated) essential singularity.

If $z = a$ is an isolated singularity,

$$\text{Res}\, f(a) = \frac{1}{2\pi i} \oint_{|z-a|=\epsilon} f(z)\, dz = c_{-1}. \qquad (7.47)$$

This has already been discussed.

Let us examine a few examples.

Example 7.7. The function, which has $z = 2$ as a first-order pole,

$$f(z) = \frac{1}{2-z} \tag{7.48}$$

is expanded as follows when $|z| < 2$:

$$f(z) = \frac{1}{2(1 - z/2)} = \frac{1}{2} \sum_{n=0}^{\infty} \left(\frac{z}{2}\right)^n. \tag{7.49}$$

On the other hand, when $|z| > 2$:

$$f(z) = -\frac{1}{z(1 - 2/z)} = -\frac{1}{z} \sum_{n=0}^{\infty} \left(\frac{2}{z}\right)^n. \tag{7.50}$$

This is a Laurent expansion around $z = 0$ when $|z| > 2$. But a Laurent expansion (7.50) does not converge when $|z| < 2$. As in this example, we can not conclude that $z = 0$ is an essential singularity just because the principal part goes on infinitely.

Example 7.8.

$$e^{1/z}. \tag{7.51}$$

$z = 0$ is an (isolated) essential singularity, and

$$e^{1/z} = 1 + \frac{1}{1!}\frac{1}{z} + \frac{1}{2!}\left(\frac{1}{z}\right)^2 + \cdots + \frac{1}{n!}\left(\frac{1}{z}\right)^n + \cdots \tag{7.52}$$

is the Laurent expansion of this function. This form clearly shows that the residue at $z = 0$ is 1.

Chapter 8

Advanced Complex Integrals

In the preceding chapters, we studied the definition of complex functions, differentiation, and integration from the perspective of calculations and the overall understanding of the subjects. In this chapter, we change a viewpoint a little from a general mathematical perspective, to see these matters again. We begin by explaining how modern mathematics examines, from the perspective of identity (equivalence), the complex space, and mapping.

8.1 Topology and topological space

8.1.1 Topology

The issue here is to understand a path on Gauss plane or the structure of Riemann surface, from a broader perspective.

The definition of distance has been introduced on Gauss plane. This kind of space is generally called a metric space. Even in an abstract space (set) in which distance is not defined, adding information on the relationship between elements can be used to proceed with a similar discussion as is possible with a metric space. This is called "defining a topology" on a space, and that space with a topology is called a "topological space".

Definition 8.1. (Topological space) The set $\mathcal{Y}$, comprising subsets Y_1, Y_2, $\cdots$ of $\mathcal{X}$, is called a family of subsets of $\mathcal{X}$. A given subset family $\mathcal{Y}$ is called a topology of $\mathcal{X}$ when $\mathcal{Y}$ satisfies the following three conditions, and $(\mathcal{X}, \mathcal{Y})$ is called a topological space[1]:

(a) The empty set $\varnothing$ (or written as { }) and the universal set $\mathcal{X}$ itself belong to $\mathcal{Y}$.

---

[1] In the following, we will refer to $\mathcal{X}$ as a topological space, *i.e.* a space defining a topology in it.

(b) The intersection $Y_1 \cap Y_2$ belongs to $\mathcal{Y}$, when subsets Y_1 and Y_2 belong to $\mathcal{Y}$.

(c) When any arbitrary number (possibly infinite) of subsets belonging to $\mathcal{Y}$ are taken, their union also belongs to $\mathcal{Y}$.

Example 8.1. (Topological space) In the following, an empty set is written as { }.

(a) If the set $\mathcal{X} = \{1, 2, 3, 4\}$ and the subset family is $\mathcal{Y} = \{\{ \}, \{1, 2, 3, 4\}\}$, this forms a topology.

(b) If the set $\mathcal{X} = \{1, 2, 3, 4\}$ and the subset family is $\mathcal{Y} = \{\{ \}, \{2\}, \{1, 2\}, \{2, 3\}, \{1, 2, 3\}, \{1, 2, 3, 4\}\}$, this forms a topology.

(c) If the set $\mathcal{X} = \{1, 2, 3, 4\}$ and the subset family is $\mathcal{Y} = \{\{ \}, \{2\}, \{3\}\}$, this does not form a topology.

Each can be directly checked for whether or not it defines a topology by Definition 8.1. We will leave this as an exercise for readers.

Example 8.2. (Structure of topological space) Let us consider a simple example. What is the space of freedom of motion corresponding to some dynamical systems in the following two-dimensional space?

(a) The simple pendulum on a plane: The degree of freedom is the angle θ ($0 \leq \theta \leq 2\pi$) of the thread that the pendulum sways.

(b) The double pendulum on a plane: The degrees of freedom are the angle θ_1 of the first weight as seen from the fixed support of the thread, and the angle θ_2 of the second weight as seen from the first weight ($0 \leq \theta_1, \theta_2 \leq 2\pi$).

In example (a), θ is a point on the two-dimensional circle (S^1), so the space is S^1. In example (b), θ_1 and θ_2 each move on independent two-dimensional circle S^1, so the space of the variable $s(\theta_1, \theta_2)$ is $S^1 \times S^1$, which is to say, the torus T^2. (The symbol $\times$ means the direct product of spaces.)

Figure 8.1 shows four types of two-dimensional surface. They are the unit circle S^1, the spherical surface S^2, the cylindrical face, and the torus T^2. On a spherical surface, any kind of non-intersecting closed path can be continuously transformed into one point. Two types of paths exist on a cylinder (of infinite length). On a torus, A or B shown in Fig. 8.1(d) cannot be continuously transformed into one point, and the continuous transformation from A to B and from B to A is also impossible.

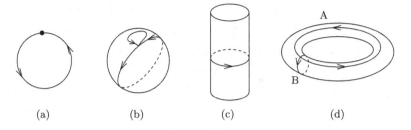

Fig. 8.1   Unit circle (a), spherical surface (b), cylinder (c), and torus (d).

Let us define a neighborhood based on topological space.

Definition 8.2. (Neighborhood) For each point (element) x in the set $\mathcal{X}$, when a family of subsets $\mathcal{U}(x)$ of $\mathcal{X}$ has the following properties, the set belonging to $\mathcal{U}(x)$ is called the **neighborhood** of x.

(a) If $U \in \mathcal{U}(x)$, then $x \in U$.
(b) If $U_1, U_2 \in \mathcal{U}(x)$, then $U_1 \cap U_2 \in \mathcal{U}(x)$.
(c) If $U \in \mathcal{U}(x)$ and $y \in U$, then $U \in \mathcal{U}(y)$.

If the neighborhood of each point x in the set $\mathcal{X}$ is included in $\mathcal{X}$, then x is called an **interior point** of $\mathcal{X}$. When all elements (points) in the set $\mathcal{X}$ are interior points, $\mathcal{X}$ is called an open set.[2] Furthermore, letting $\mathcal{X}^c$ be the complement of $\mathcal{X}$, when any neighborhood $\mathcal{U}$ of a point $a \in \mathcal{X}$ satisfies conditions $\mathcal{X} \cap \mathcal{U} \neq \varnothing$ and $\mathcal{X}^c \cap \mathcal{U} \neq \varnothing$, the point a is called the boundary point of $\mathcal{X}$. The set of all boundary points of $\mathcal{X}$ is called the **boundary** of $\mathcal{X}$.

Definition 8.3. (ε neighborhood) When the distance $\rho(a, x)$ between any arbitrary points a and x in the set $\mathcal{X}$ is defined, this set is called a metric space. For a point a and a positive number $\varepsilon > 0$, the subset $U_\varepsilon(a) = \{x \in \mathcal{X} | \rho(a, x) < \varepsilon\}$ of $\mathcal{X}$ is called the **ε neighborhood** of a.

The approach of topology allows us to give a general definition of the **continuity of functions.**[3]

---

[2]For the empty set $\varnothing$, the assumption "if $x \in \varnothing$" is not true, so this is understood as being included in the conditions for the statement "all points included in the empty set $\varnothing$ are interior points", which is used in the definition of an interior point, to be valid, and the empty set $\varnothing$ is understood to be an open set.

[3]**[Definition of a continuous function by employing ε-δ]**
For all points z inside a circle of a radius δ centered at the point z_0 on a Gauss plane

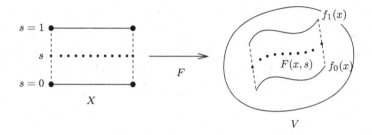

Fig. 8.2   Continuous mapping F.

Definition 8.4. (Definition of continuous mapping, based on the concept of topological space) Let $f : X \to Y$ be a mapping from a topological space X to a topological space Y. The fact that the mapping f is continuous at $x \in X$ means that for any neighborhood V of $f(x) \in Y$, the image of the neighborhood can be $f(U(x)) \subseteq V$, if we take an appropriate neighborhood $U(x)$ of x. In other words, the inverse image $f^{-1}(V)$ of the set V is an open set.

8.1.2   Equivalence relation of continuous mapping

Our problem is to know the nature of the mapping.

Let f_0 and f_1 be a continuous mapping from topological space X to topological space V.

$$f_0,\ f_1\ :\ X \to V \tag{8.1}$$

In this case, if there exists a continuous transformation from f_0 to f_1, it follows that f_0 and f_1 are of the same type (Fig. 8.2). The following is a slightly more rigorous statement.

Definition 8.5. (Equivalence, homotopic) X and V are topological spaces, and f_0 and f_1 are two continuous mappings from X to V. For a mapping F from product space $X \times [0,1]$ to space V

$$F\ :\ X \times [0,1] \to V, \tag{8.2}$$

if F is a continuous mapping that satisfies

$$F(x,0) = f_0(x), \qquad F(x,1) = f_1(x), \tag{8.3}$$

---

and for an arbitrarily small positive number ε, if

$$|f(z) - f(z_0)| < \varepsilon,$$

the complex function $f(z)$ is said to be continuous at z_0.

f_0 and f_1 are said to be homotopically equivalent or homotopic, and this relation is written as

$$f_0 \sim f_1. \tag{8.4}$$

A homotopic relation $\sim$ is an equivalence relation. That is

- $f \sim f$,
- If $f \sim g$, then $g \sim f$,
- If $f \sim g$ and $g \sim h$, then $f \sim h$.

8.1.3 Loops and fundamental groups

Let us define a path with an initial point x_0 and a final point x_1 in space V. The parameter t $(0 \leq t \leq 1)$ may be used to express the path α as

$$\alpha(t): \quad \alpha(0) = x_0, \ \alpha(1) = x_1. \tag{8.5}$$

Figure 8.3 depicts two types of space on a two-dimensional plane, and paths on them. (a) is a space with no holes, while (b) is a space with a hole. Looking at the figure, it is easy to distinguish between the two types of space.

In space V, a closed path that starts from one point x_0 and ends at the same x_0 (the initial point and final point are the same) is called a **closed path** or **loop** with the **base point** x_0. Let us try continuously transforming the loop drawn in the space. In a space with no hole, it is possible to continuously reduce any arbitrary loop to one point. In a space with a hole, it is possible to reduce a loop to one point, if the loop does not encircle the hole. If, however, the loop does encircle the hole, the

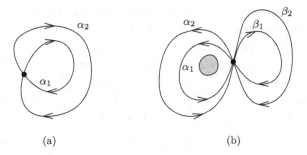

(a)   (b)

Fig. 8.3   (a) A space with no holes, and (b) a space with a hole, on a two-dimensional plane. In space (a), loops α_1 and α_2 are homotopic $(\alpha_1 \sim \alpha_2)$. In space (b), $\alpha_1 \sim \alpha_2$, and $\beta_1 \sim \beta_2$, but $\alpha_1 \not\sim \beta_1$.

path cannot jump over the hole to reduce to one point by any continuous transformation.

Thus, in space (a) with no holes, there is only one type of loop, but in space (b) with a hole, there are two types of loops. In space with a hole, one type is that the loop can be reduced to a single point by continuous transformation as in (a). The other type is that the loop cannot be reduced to one point by continuous transformation. These properties of loops are unique ones of a given space and are not changeable.

In a space (a) with no hole, all loops are contractable to a single point, so there is only one class of loop, called **homotopy class**. In the following discussion, we will refer to a 'path' that is contacted to one point as a "zero loop". In a space with no holes, therefore, all loops are contractable to a zero loop. On the other hand, in a space (b) with one hole, there is an infinite number of homotopy classes, corresponding to loops that do not rotate around a hole (and are contactable to a point) and those which rotate n times clockwise or counterclockwise around the hole. Thus, the properties of each type of space can be characterized by investigating the homotopy class.

8.1.4 *Homotopy and products of loops*

To construct algebraic structure (where one should define "product" and "inverse"), we must first define the "product" of paths. When the final point of $f_1(t)$ and the initial point of $f_2(t)$ match for the paths $f_1(t)$ and $f_2(t)$ (*i.e.* $f_1(1) = f_2(0)$), connecting one end of a path to one end of the other path to form one path, one can define the product of two paths as "product $(f_1 \cdot f_2)$". (The product cannot be defined for two paths when the end point of one and the start point of the other do not match.) For the path f_1, the inverse f_1^{-1} is defined as the path which proceeds backward along f_1 from the final point to the initial point. A loop is a special path in which the initial point and final point are the same, and, in defining the product and the inverse, we do not need any additional statements.

Consider homotopic loops $f_0(t)$ and $f_1(t)$ ($f_0 \sim f_1$), with the base point x_0. The loop $f_u(t)$ ($0 < u < 1$), which is homotopic with these loops, also exists. Therefore, we can consider a continuous mapping

$$F : I \times I \to V \qquad (8.6)$$

which satisfies

$$F(t, 0) = f_0(t), \quad F(t, 1) = f_1(t), \quad F(0, u) = F(1, u) = x_0. \qquad (8.7)$$

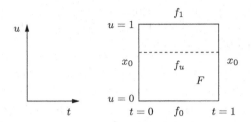

Fig. 8.4 Mapping $F : I \times I \to V$. F is expressed as a square. A variation on the horizontal axis represents parameter t, which expresses the movement of a point on a single path. A variation on the vertical axis represents the variation of parameter u, which expresses the transformation of the path. The bottom represents f_0, the top represents f_1, and the left and right vertical sides represent the base point x_0. The first $I = [0, 1]$ of $I \times I$ represents the range of t, and the second I represents the range of u.

Figure 8.4 shows the mapping F. This mapping F is called a **homotopy** between f_0 and f_1.

If $f \sim g$ and $0 \leq t \leq 1$, the product $f \cdot g$ of the two loops is defined as

$$(f \cdot g)(t) = \begin{cases} f(2t) & (0 \leq t \leq 1/2) \\ g(2t - 1) & (1/2 \leq t \leq 1) \end{cases}. \qquad (8.8)$$

From Fig. 8.4 and this definition, one can see

$$f_0 \cdot g_0 \sim f_1 \cdot g_1 \quad \text{if} \quad f_0 \sim f_1 \quad \text{and} \quad g_0 \sim g_1. \qquad (8.9)$$

The outline of the proof of Eq. (8.9) is shown in Fig. 8.5. Furthermore, the inverse of loop $f(t)$ can be defined as moving around the loop inversely, which is to say, it can be considered to be $f^{-1}(t) = f(1 - t)$.

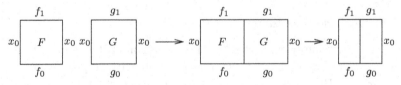

Fig. 8.5 The proof that if $f_0 \sim f_1$ and $g_0 \sim g_1$, then $f_0 \cdot g_0 \sim f_1 \cdot g_1$.

By defining product and inverse as above, the "equivalence class" of loops with the base point x_0 (the set of all elements which are homotopic with an element f) has the structure of a group.[4] This group is called

---

[4]In the set $\mathcal{G}$, having elements $a_0, a_1, a_2, \cdots$, a unique element $\mathcal{G}$, called the product of any two arbitrary elements a_i and a_j is defined (written as $a_i \cdot a_j$), and the set $\mathcal{G}$ is called a group when the following three properties are satisfied:
- Associative: $a_i \cdot (a_j \cdot a_k) = (a_i \cdot a_j) \cdot a_k$

fundamental group, and is written as $\pi_1(V, x_0)$. A fundamental group is also called a one-dimensional **homotopic group**. We will not discuss higher-order homotopic groups here.

8.1.4.1 *Replacing base point*

The fundamental group $\pi_1(V, x_0)$ is defined with the base point x_0 fixed so far. However, it is not essential to fix the base point. Let the path g have the initial point x_1 and the final point x_2. If loop f has x_2 as its base point $g \cdot f \cdot g^{-1}$ is the loop having x_1 as its base point. In a space where a path such as g can certainly be drawn, it is always possible to replace the base point as above. Therefore, this group can be regarded as a group determined by V alone, without depending on how the base point is set. This group is expressed as $\pi_1(V)$, and is called the fundamental group of V.

Let us examine fundamental groups by following some examples.

Example 8.3. (Example of fundamental group)

(a) All loops on a spherical surface S^2 can contract to a single point, so $\pi_1(S^0) = 0$. This is the only homotopy class. In this space, the fundamental group comprises only one element, which is to say, only an identity element.

(b) For a loop on the unit circle S^1, $\pi_1(S^1) = \mathbb{Z}$ (where $\mathbb{Z}$ is the group consisting of positive and negative integers and 0). A loop on a circle goes around n times either counterclockwise or clockwise or does "not at all", corresponding to positive and negative numbers and zero.

8.1.4.2 *Simply-connected*

We discussed "simply-connected" in the discussion of Riemann surface. In that discussion, we understood it to be a domain with no poles or no other singularities. Here, we will redefine this idea using the term "topological space".

Definition 8.6. (Simply connected) It is assumed that any two points in a topological space V are connected by a continuous path, and any loop with arbitrary base point $x_0 \in V$ is homotopic to a zero loop. In this case, the space V is said to be **simply-connected**.

- Identity element a_0: A unique a_0 exists that satisfies $a_i \cdot a_0 = a_0 \cdot a_i = a_i$.
- Inverse element $(a_i)^{-1}$ of the element a_i: An $(a_i)^{-1}$ exists that satisfies $(a_i)^{-1} \cdot a_i = a_i \cdot (a_i)^{-1} = a_0$.

As this definition makes clear, the spherical surface S^2 is simply-connected, but the circle S^1 is not.

8.1.5 *Degree or rotation number*

Let us consider the problem of how many times the image of the continuous mapping f on the circle S^1 ($f : S^1 \to S^1$) covers S^1. Consider the circle S^1 as a set of complex numbers z of an absolute value 1 ($S^1 = \{ e^{i\theta} \mid \theta \in \mathbb{R} \}$, $\mathbb{R}$ is the set of real numbers). Then the image of the mapping z^k covers S^1 k times. When the mapping f covers S^1 k times in this way, k is called the **degree**, or the **rotation number**. It is known that the degree k does not change if f transforms continuously. Therefore, the degree is an important quantity (**homotopy invariance**) characterizing the mapping.

Let us discuss this in more detail. f is any arbitrary continuous mapping that satisfies

$$f : S^1 \to S^1.$$

Here, we will define another continuous mapping $\phi : \mathbb{R} \to S^1$ as

$$\phi(\theta) = e^{i\theta}. \tag{8.10}$$

It is possible to choose a continuous mapping f_* for an arbitrary f such as

$$\phi \cdot f_* = f \cdot \phi \qquad (\mathbb{R} \to S^1). \tag{8.11}$$

Rewriting Eq. (8.11) in more concretely, we get

$$e^{i f_*(\theta)} = f(e^{i\theta}). \tag{8.12}$$

Thus,

$$f_*(\theta) = \arg f(e^{i\theta}) - i \ln |f(e^{i\theta})|. \tag{8.13}$$

Mapping f is $S^1 \to S^1$, so when θ rotates once around the circle S^1, in general $f(e^{i\theta})$ rotates several times around S^1. With Eq. (8.11), one can write

$$\phi(f_*(2\pi)) = f(\phi(2\pi)) = f(e^{i2\pi}) = f(\phi(0)) = \phi(f_*(0)). \tag{8.14}$$

From this,

$$f_*(2\pi) - f_*(0) = 2m\pi, \tag{8.15}$$

where m is an integer. The argument $\arg z$ of a complex number z has an indefiniteness of $2n\pi$, but taking the difference as in Eq. (8.15), m does not depend on how the argument of Eq. (8.13) is chosen. This m is called the "degree" or the "rotation number" of f, written as $\deg f$.

We obtain the following two theorems. Intuitively, they can be understood from the explanation so far, but readers may refer to other sources for concrete proof.

Theorem 8.1. *Following equation holds for the two continuous mappings* f, g: $S^1 \to S^1$,

$$\deg(f \cdot g) = \deg f \cdot \deg g. \tag{8.16}$$

Theorem 8.2. *Necessary and sufficient conditions for two continuous mappings* f, g: $S^1 \to S^1$ *to be* $f \sim g$ *is*

$$\deg f = \deg g. \tag{8.17}$$

8.2 Index of a point concerning closed curve, and generalized residue theorem

8.2.1 *Index of a point concerning closed curve (winding number)*

Before generalizing Cauchy's theorem, let us determine how many times the closed curve C encircles a point z_0 which is not on that curve C. First, we define

$$n(C, z_0) = \frac{1}{2\pi i} \int_C \frac{dz}{z - z_0}, \tag{8.18}$$

which is called the **index**, or the **winding number**, of the closed curve C concerning the point z_0.

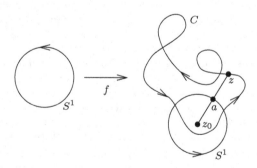

Fig. 8.6 Composite map $g \cdot f : S^1 \to C \to S^1$ and the definition of the index.

Suppose that the mapping f is a continuous mapping from S^1 to the closed curve C (Fig. 8.6)

$$f : S^1 \to C. \tag{8.19}$$

Choose a point z_0, not on the curve C, and consider a unit circle S^1 whose center is z_0. Draw a line from z_0 to the point z on curve C and let the intersection between the half line and the unit circle S^1 be a. Let this correspondence $(z \to a)$ be the mapping g.

$$g : C \to S^1. \tag{8.20}$$

The composite map $g \cdot f$ of f and g defined thus is

$$g \cdot f : S^1 \to S^1. \tag{8.21}$$

The degree $\deg(g \cdot f)$ of $g \cdot f$ is equivalent to the definition of the index in Eq. (8.18). This fact can be seen in Fig. 8.6. Let us examine a few examples.

Example 8.4. (Examples of index)

(a) The closed curve C is a unit circle with the origin as its center, going around in a counterclockwise direction. The index is 1 if the point z_0 is inside the unit circle, and 0 if it is outside.

(b) As the closed curve C, consider a path that rotates clockwise along a unit circle with the origin as its center. The index is -1 if the point z_0 is inside the unit circle, and 0 if it is outside.

(c) In an example shown in Fig. 8.7, the index is 1 if point z_0 is in one of the areas labeled A, it is 2 if z_0 is in B, and 0 if it is in other domains outside.

For the sake of brevity, we now introduce the concept of homology equivalence, using the index of a curve.

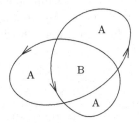

Fig. 8.7   Index for Example 8.4(c).

Definition 8.7. For all points $a \notin D$, when the closed curve C in an open domain D satisfies $n(C, a) = 0$, the closed curve C is said to be **homologous to 0**.

For the closed curves C_1 and C_2, when $C_1 - C_2$ is inside the domain D,[5] and $C_1 - C_2$ is homologous to 0, C_1 and C_2 are **homologous** in D. This means that, for all points $a \notin D$,

$$n(C_1, a) = n(C_2, a) \tag{8.22}$$

is true. If they are homotopic, they are homologous, but the converse is not always true (Fig. 8.8).

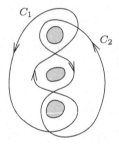

Fig. 8.8   Two closed curves C_1 and C_2 are homological equivalent, since both two curves go around the first and third holes counterclockwise, but do not go around the second hole. However, C_1 and C_2 are not homotopic.

8.2.2   *Generalization of Cauchy's integral formula*

Consider the complex function $f(z)$, which is holomorphic in an open domain D. We have already presented **Cauchy's integral formula** for the holomorphic function f in Chapter 7.

Example 8.5. (Cauchy's integral formula) Let D be a simply-connected domain in which the complex function $f(z)$ is holomorphic, and C' be a contour path traversed in the positive direction on a closed Jordan curve (a closed curve that does not intersect with itself) in D. For any arbitrary point z interior to C',

$$f(z) = \oint_{C'} \frac{f(\zeta)}{\zeta - z} \, d\zeta.$$

---

[5]When written in this way, it means that for the closed curves C_1 and C_2, $C_1 + (-C_2)$ is a single closed curve. Therefore, the assumptions are that $(-C_2)$ is the path that follows C_2 in the reversed direction, and that C_1 and C_2 have the same base point.

This is Cauchy's integral formula Eq. (7.1) as we have considered it until now.

From now on, we will consider providing integral formulae in a general form that includes Cauchy's integral formula. Let $f(z)$ be holomorphic in the domain D, and consider

$$F(z) = \frac{f(z) - f(z_0)}{z - z_0}. \tag{8.23}$$

This function is holomorphic if $z \neq z_0$, and z_0 is a removable singularity of $F(z)$. In this case, integrating around the (closed) path C, which is homologous to 0 inside this domain, we get

$$\int_C F(z)\,\mathrm{d}z = 0, \tag{8.24}$$

(cf. Sec. 6.4.1, Cauchy's integral theorem). The point z_0, in this case, can be either interior or exterior to the closed curve C. Equation (8.24) is rewritten, with Eq. (8.23), as

$$\int_C \frac{f(z)}{z - z_0}\,\mathrm{d}z = f(z_0) \int_C \frac{1}{z - z_0}\,\mathrm{d}z. \tag{8.25}$$

The integration of the right-hand side is "the index of the closed curve C concerning the point z_0" $\times\ 2\pi\mathrm{i}$. The above can be summarized as follows.

Theorem 8.3. (Generalization of Cauchy's integral formula) *Suppose that $f(z)$ is holomorphic in the open domain D, and that C is a closed curve, homologous to 0, in D. In this case, for any arbitrary point z_0 that is not on C, we get*

$$n(C, z_0) \cdot f(z_0) = \frac{1}{2\pi\mathrm{i}} \int_C \frac{f(z)}{z - z_0}\,\mathrm{d}z\ . \tag{8.26}$$

Here, $n(C, z_0)$ is the index of the closed curve C concerning the point z_0 (the winding number of C around z_0).

Under this theorem, point z_0 can be outside the holomorphic domain of $f(z)$, because, for such z_0, $n(C, z_0) = 0$. The case of $n(C, z_0) = 1$ is Cauchy's integral formula stated in Sec. 7.1.1.

8.2.3 Generalization of residue theorem

The residue theorem (Th. 6.7) can be stated generally, in the same way as the general form of Cauchy's integral formula (8.26).

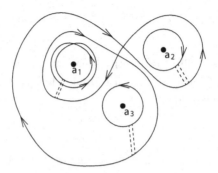

Fig. 8.9   Generalization of the residue theorem.

Theorem 8.4. (Generalization of residue theorem) *Suppose that the function f is holomorphic in the open domain D, except a finite number of singularities* $a_1, a_2, \cdots, a_n$, *and the residue of the function f at* a_j *is* $\mathrm{Res}\,(f, a_j)$. *If we further suppose that the closed curve C is homologous to 0 in the domain D, and that each point* a_j *is not on C, then we get*

$$\frac{1}{2\pi i} \int_C f(z)\, dz = \sum_j n(C, a_j)\, \mathrm{Res}\,(f, a_j). \qquad (8.27)$$

Proof. Points a_j with $n(C, a_j) = 0$ can be excluded from the discussion. Assume that a path C_j is a small circle with a center a_j, traversed in the positive direction (counterclockwise) and, therefore, C_j does not overlap with C. We write a path, if this contour path goes n times around a_j on C_j, as nC_j. Suppose that the closed curve C does not cross C_j nor its interior, and let it be included in D and in the open domain $D - \{a_j\}$. Once we add a small supplementary path going back and forth along the same route and link the closed curve C and $-n(C, a_j)C_j$ together, we get a closed curve $C' = C - \sum n(C, a_j)C_j$ homologous to 0 on the open domain $D - \{a_j\}$ (Fig. 8.9). Rewriting by applying Cauchy's integral theorem (8.24) to this curve C', we get

$$\frac{1}{2\pi i} \int_C f(z)\, dz = \sum_j n(C, a_j) \frac{1}{2\pi i} \int_{C_j} f(z)\, dz = \sum_j n(C, a_j)\, \mathrm{Res}\,(f, a_j).$$

This is Eq. (8.27).   □

8.3   Application of residue theorem: Argument principle and Rouché's theorem

We will now describe some important applications of the residue theorem.

8.3.1 The orders of poles and zero points

We have mentioned the concept of the order of a pole before, but we did not pay any particular attention to zero points. First, let us define several terms.

Definition 8.8. (Order of pole) When, in the neighborhood of $z = z_0$ excluding z_0, the complex function $f(z)$ is

$$f(z) = \frac{a_{-k}}{(z - z_0)^k} + \frac{a_{-k+1}}{(z - z_0)^{k-1}} + \cdots + \frac{a_{-1}}{(z - z_0)} + a_0 + a_1(z - z_0) + \cdots , \quad (8.28)$$

k is called the **order** of the pole z_0.

This was described in Chapter 5. The part of negative power is called the **principal part** (Sec. 7.3.2).

Definition 8.9. (Meromorphic) A complex function that has no singularities other than poles, at most, in the domain D is said to be meromorphic in D.[6]

Theorem 8.5. *A complex function that has no singularities other than poles (it has no essential singularity), at most, on the entire z plane including $z = \infty$, is nothing more than a rational function.*

Proof. Consider a (meromorphic) complex function $f(z)$ that has no singularities other than poles within a finite range from the origin. Let p poles in the finite domain of $f(z)$ be $\beta_1, \beta_2, \cdots, \beta_p$, and their corresponding orders be $k_1, k_2, \cdots, k_p$. Write the principal part of $f(z)$ at the pole β_i (with order k_i) as

$$H_i(z - \beta_i) = \frac{a_{-k_i}}{(z - \beta_i)^{k_i}} + \frac{a_{-k_i+1}}{(z - \beta_i)^{k_i-1}} + \cdots + \frac{a_{-1}}{(z - \beta_i)}.$$

Writing

$$g(z) = f(z) - \sum_{i=1}^{p} H_i(z - \beta_i), \quad (8.29)$$

$g(z)$ is holomorphic in the finite range of z. Also, because there is no pole, at most, except $z = \infty$, $g(z)$ is meromorphic. Therefore, if $g(z)$ is

---

[6]The concept of a rational function differs from that of a meromorphic function. The rational function

$$f(z) = \frac{a_0 z^n + a_1 z^{n-1} + \cdots + a_{n-1} + a_n}{b_0 z^m + b_1 z^{m-1} + \cdots + b_{m-1} z + b_m}$$

was described in Chapter 4.

holomorphic at $z = \infty$, it is a constant according to Liouville's theorem. Also, if $z = \infty$ is a pole of order q, its principal part is

$$H_\infty(z) = c_1 z + c_2 z^2 + \cdots + c_q z^q \qquad (c_q \neq 0).$$

Then, since $g(z) - H_\infty(z)$ is holomorphic in entire domain, including $z = \infty$, $g - H_\infty = c_0$ (constant). Therefore,

$$f(z) = g(z) + \sum_{i=1}^{p} H_i(z - \beta_i)$$

$$= c_0 + c_1 z + c_2 z^2 + \cdots + c_q z^q + \sum_{i=1}^{p} H_i(z - \beta_i), \qquad (8.30)$$

and this is a rational function. □

Definition 8.10. (Order of zero point) Suppose that the function $f(z)$ is holomorphic in the domain D, and that $f(z_0) = 0$ at the point z_0 in D. z_0 is said to be a **zero point** (or simply "zero") of the holomorphic function $f(z)$. The name "zero point" has been used a few times by now. If z_0 is a zero point, $f(z)$ can be expanded into a Taylor series at $z = z_0$. When the Taylor expansion is

$$f(z) = c_k(z - z_0)^k + c_{k+1}(z - z_0)^{k+1} + c_{k+2}(z - z_0)^{k+2} + \cdots, \qquad (c_k \neq 0)$$
$$(8.31)$$

k is called the **order** of the zero point z_0 of $f(z)$.

8.3.2 *Argument principle*

We will now give proof for the "argument principle", first with the concept discussed before. After that, we will explain this with the new concept "the index of a point".

Theorem 8.6. (Argument principle I) *Suppose the function $f(z)$ is meromorphic in the simply-connected domain D, and there are no zero points or poles of $f(z)$ on the closed Jordan curve C (a closed curve that does not intersect with itself) in D. If the zero points and poles of $f(z)$ inside C (their corresponding orders) are as follows;*

zero points $a_1, a_2, \cdots,$ *with the orders* $h_1, h_2, \cdots,$

poles $b_1, b_2, \cdots,$ *with the orders* $k_1, k_2, \cdots,$

then

$$\frac{1}{2\pi i} \oint_C \frac{f'(z)}{f(z)} dz = \frac{1}{2\pi} \oint_C d(\arg f(z)) = \sum_j h_j - \sum_l k_l. \qquad (8.32)$$

Proof. If $z = a$ is a zero point or a pole of $f(z)$, then $z = a$ is a first order pole of $f'(z)/f(z)$.

First, if $z = a$ is an order-h zero point, then

$$f(z) = (z - a)^h g_1(z), \qquad (8.33a)$$

where $z = a$ is a holomorphic point and not a zero point of $g_1(z)$. Therefore,

$$\frac{f'(z)}{f(z)} = \frac{h}{z - a} + \frac{g_1'(z)}{g_1(z)}, \qquad (8.33b)$$

and $z = a$ is a holomorphic point of g_1'/g_1. If $z = b$ is an order-k pole of $f(z)$, we can write

$$f(z) = (z - b)^{-k} g_2(z) \qquad (8.34a)$$

and $z = b$ is a holomorphic point and not a zero point of $g_2(z)$. Therefore,

$$\frac{f'(z)}{f(z)} = -\frac{k}{z - b} + \frac{g_2'(z)}{g_2(z)} \qquad (8.34b)$$

and $z = b$ is a holomorphic point of g_2'/g_2. From Eqs. (8.33b) and (8.34b), we get directly

$$\oint_C \frac{f'(z)}{f(z)} dz = 2\pi i \sum_j h_j - 2\pi i \sum_l k_l. \qquad (8.35)$$

The right-hand side of Eq. (8.35) is determined by the order of the poles and zero points included inside the closed curve C. $\qquad \square$

$f'(z)/f(z)$ in the left-hand side of Eq. (8.32) can be transformed to

$$\frac{f'(z)}{f(z)} dz = d(\log f(z)) = d \ln |f(z)| + i\, d(\arg f(z)). \qquad (8.36)$$

From this, it follows that

$$\frac{1}{2\pi i} \oint_C \frac{f'(z)}{f(z)} dz = \frac{1}{2\pi i} \oint_C d(\ln |f(z)|) + \frac{1}{2\pi} \oint_C d(\arg f(z)). \qquad (8.37)$$

Since $\ln |f(z)|$ is a single-valued function, the first term on the right-hand side becomes 0 in the integration on a closed curve and

$$\frac{1}{2\pi i} \oint_C \frac{f'(z)}{f(z)} dz = \frac{1}{2\pi} \oint_C d(\arg f(z)). \qquad (8.38)$$

This produces a relation between the sum (left-hand side) of the orders of zero points and poles inside a closed path C of $f(z)$, and the variation of the argument$/2\pi$ along the closed curve C, which is to say, the index (winding number). This is the reason why we call Th. 8.6 the argument principle.

We can present a generalization of the argument principle with the idea of the index of the closed path.

Theorem 8.7. (Argument principle II) *Suppose that in an open domain D there is a non-constant function $f(z)$ which has, at most, a finite number of zero points and poles, and is holomorphic in D except these poles. Furthermore, suppose that a closed path C is homologous to 0 in the domain D and that the zero points and poles of f do not locate on C. If the zero points and poles of $f(z)$ that are inside C, and their corresponding orders, are*

> zero points $a_1, a_2, \cdots,$ with the orders $h_1, h_2, \cdots,$
>
> poles $b_1, b_2, \cdots,$ with the orders $k_1, k_2, \cdots,$

then, using the indices of the zero point a_j and the pole b_l, the integration of $f'(z)/f(z)$ along C can be written as

$$\frac{1}{2\pi i} \int_C \frac{f'(z)}{f(z)}\, dz = \sum_j n(C, a_j)\, h_j - \sum_l n(C, b_l)\, k_l. \qquad (8.39)$$

This theorem can be proved, by combining the idea of indices and the proof of Eq. (8.32), and we omit it here.

8.3.3   *Rouché's theorem*

Rouché's theorem is closely related to the argument principle.

8.3.3.1   *Rouché's theorem*

Theorem 8.8. (Rouché's theorem) *Consider a closed curve C in a simply-connected holomorphic domain D of $f(z)$, $g(z)$. Suppose that there are no zero points of $f(z)$, $g(z)$ on the curve C. In this case, if on the curve C*

$$|f(z)| > |g(z)| \qquad (8.40)$$

holds, $f(z)$ and $f(z) + g(z)$ have the same number of zero points inside C, counting duplicates order times.

Proof. Since $|f(z)| > |g(z)|$, $|f + g| \geqq |f| - |g| \geqq 0$. $f(z)$ and $f(z) + g(z)$ have no poles in D. Therefore, by the argument principle, a difference of numbers of zero point between them is

$$\frac{1}{2\pi} \oint_C d(\arg(f(z) + g(z))) - \frac{1}{2\pi} \oint_C d(\arg(f(x)))$$

$$= \frac{1}{2\pi} \oint_C d(\arg(1 + g(z)/f(z))).$$

Because of the assumption $|f| > |g|$, the image of C by $F(z) = 1 + g/f$ is inside the unit circle centered at $F = 1$. In other words, the image of C by $F(z)$ does not encircle around the origin $F = 0$. Therefore,

$$\oint_C dz \arg(1 + g(z)/f(z)) = 0.$$

Then the numbers of zero points of $g(z)$ and $f(z) + g(z)$ are equal, counting duplicates order-times. □

8.3.3.2 *Fundamental theorem of algebra*

Before we used Liouville's theorem to prove the fundamental theorem of algebra. Here, we will present a proof using Rouché's theorem.

Theorem 8.9. (Fundamental theorem of algebra)

$$f(z) = z^n + a_1 z^{n-1} + \cdots + a_{n-1} z + a_n = 0 \qquad (n \geq 1) \qquad (8.41)$$

has n roots (zero points), counting duplicates order-times, of complex numbers.

Proof. If a sufficiently large value of R is taken, it produces

$$1 + |a_1| + \cdots + |a_{n-1}| + |a_n| \leq R. \qquad (8.42)$$

At the point z on a circle with radius R chosen in that way,

$$\begin{aligned}
|f(z) - z^n| &= |a_1 z^{n-1} + a_2 z^{n-2} + \cdots + a_{n-1} z + a_n| \\
&\leq |a_1| R^{n-1} + |a_2| R^{n-2} + \cdots + |a_{n-1}| R + |a_n| \\
&\leq |a_1| R^{n-1} + |a_2| R^{n-1} + \cdots + |a_{n-1}| R^{n-1} + |a_n| R^{n-1} \\
&\leq R^n = |z^n|. \qquad (8.43)
\end{aligned}$$

According to Rouché's theorem, the numbers of roots (zero points) of $f(z)$ and z^n inside a circle of sufficiently large radius R are equal, so $f(z)$ has n roots. □

Compare this proof with the one given in Sec. 7.3.1.

Chapter 9

Analytic Continuation and Riemann Surface

We will now explain "analytic continuation", which is one of the most important issues of complex analysis. In a domain in which a holomorphic function is defined, for example, on the real axis, analytic continuation provides a method to extend the range of definitions of this function to the complex plane.

For single-valued holomorphic functions, rewriting the differentiable real function $f(x)$ by replacing $x \to z$ yields the holomorphic function $f(z)$ corresponding to $f(x)$ on the real axis. We will also introduce Weierstrass' method which uses the Taylor expansion to perform analytic continuation generally. Riemann surface can be understood in more concrete terms through the analytic continuation presented. In Chapter 10, as some other examples of analytic continuation, Γ (gamma) functions and B (beta) functions will be discussed.

9.1 Theorem of identity and reflection principle

9.1.1 Theorem of identity

Theorem 9.1 (Theorem of identity). *If $f(z)$ is holomorphic in the domain D and $f(z_k) = 0$ holds at each point of a sequence $\{z_k\}$ in D, which converges to point a in D, then*

$$f(z) = 0 \quad \text{where} \quad z \in D. \tag{9.1}$$

Proof. As $f(z)$ is holomorphic in D, the Taylor expansion can be used inside a circle of radius R centered at a ($|z - a| < R$).

$$f(z) = c_0 + c_1(z - a) + c_2(z - a)^2 + \cdots \tag{9.2a}$$

151

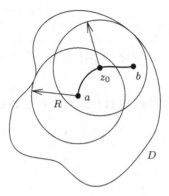

Fig. 9.1   Convergence circle $|z - a| < R$ and the connection from point a to point b.

As $f(z)$ is continuous at $z = a$,

$$c_0 = f(a) = \lim_{k \to \infty} f(z_k) = 0. \tag{9.2b}$$

Next, when

$$f_1(z) = \frac{f(z)}{z - a} = c_1 + c_2(z - a) + c_3(z - a)^2 + \cdots \tag{9.2c}$$

is defined, $f_1(z)$ is also holomorphic inside the convergence circle of the series, and $f_1(z_k) = f(z_k)/(z_k - a) = 0$, so

$$c_1 = f_1(a) = \lim_{k \to \infty} f_1(z_k) = 0. \tag{9.2d}$$

In the same way,

$$c_0 = c_1 = c_2 = \cdots = 0 \tag{9.2e}$$

and it is clear that inside the convergence circle $|z - a| < R$ centered at $z = a$, $f(z)$ is identically 0 (Fig. 9.1).

Next, we consider any arbitrary point b ($|b - a| > R$) outside the convergence circle but in D. Connect a and b with an appropriate curve $z(t)$ in D and choose z_0 in the convergence circle ($|z_0 - a| < R$) on that curve. On the curve $z(t)$, it is possible to choose a sequence of points $\{z_k'\}$ between a and z_0 which is $\lim_{k \to \infty} z_k' = z_0$ and $|\forall z_k' - a| < R$. Then expand into a Taylor series around z_0, whose convergence radius is R'. Repeat the same discussion as above, and the result tells that $f(z) = 0$ for all points in the convergence circle $|z - z_0| < R'$ centered at z_0. Repeating this process several times, we can get a Taylor series whose convergence circle contains the point b and the same discussion leads to $f(b) = 0$. $\square$

The next theorem can be derived immediately from the one above.

Theorem 9.2. *If $f_1(z)$ and $f_2(z)$ are holomorphic in the domain D, and $f_1(z_k) = f_2(z_k)$ at each point in the point sequence $\{z_k\}$ in D, which converges to a point a in D, it follows that $f_1(z) = f_2(z)$ for all points in D.*

Proof. The theorem of identity can be applied to $f(z) = f_1(z) - f_2(z)$. □

Theorem 9.3. *If $f(z)$ is holomorphic in D, and at a point a in D, if*

$$f(a) = f^{(k)}(a) = 0 \quad (k = 1, 2, \cdots) \tag{9.3a}$$

then

$$f(z) = 0, \ \forall z \in D. \tag{9.3b}$$

Proof. In a domain centered on $z = a$, $f(z)$ can be written as

$$f(z) = c_0 + c_1(z - a) + c_2(z - a)^2 + \cdots. \tag{9.4a}$$

If $f(a) = f'(a) = f^{(2)}(a) = \cdots = 0$, then

$$c_0 = c_1 = c_2 = \cdots = 0. \tag{9.4b}$$

In a small domain centered at $z = a$, $f(z) = 0$, so, according to the theorem of identity, $f(z) = 0$ everywhere in D. □

If $f_1(z) = f_2(z)$ for the functions $f_1(z)$ and $f_2(z)$ in a small area of domain D, then $f_1(z) = f_2(z)$ throughout the holomorphic domain D. This is a conclusion to easily reach by considering $f(z) = f_1(z) - f_2(z)$ and Th. 9.3.

Example 9.1. If a function $\exp x$ defined on the real axis is extended to the complex z plane, it is $\exp z$.

Example 9.2. If a function $\sin x$ defined on the real axis is extended to the complex z plane, it is $\sin z$.

9.1.2 The reflection principle

Keeping the theorem of identity in mind, let us now consider the following. Consider the domain D which contains a part of the real axis (a line segment) and a holomorphic function $f(z)$ in D. Suppose that a function $f(z)$ is real on the real axis. In this case, on the real axis $z = x$, we get

$$f(z) = f(\bar{z}) = f(x)$$
$$\overline{f(z)} = \overline{f(x)} = f(x).$$

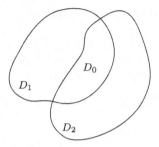

Fig. 9.2   Analytic continuation of a simply-connected domain ($D_0 = D_1 \cap D_2$).

Therefore, on the real axis in the domain, we also get

$$f(\bar{z}) - \overline{f(z)} = 0. \tag{9.5}$$

If we use the theorem of identity here, this relation can be extended to the whole domain D and we get the following theorem.

Theorem 9.4 (Reflection principle). *Suppose that a function $f(z)$ is holomorphic in the region D, which contains a part of the real axis (a line segment), and the value of $f(z)$ is real on the real axis of z plane in D. In that case, in the whole region D,*

$$f(\bar{z}) = \overline{f(z)}. \tag{9.6}$$

The reflection principle provides a way of achieving an analytic continuation for a holomorphic function defined on a part of the upper (lower) half z plane to be extended beyond the real axis to another half z plane. See, as an example, Sec. 11.1.5.

9.2   Analytic continuation and Riemann surfaces

9.2.1   *Analytic continuation*

9.2.1.1   *Analytic continuation and its uniqueness*

Suppose that the complex function $f_1(z)$ is holomorphic in a domain D_1, $f_2(z)$ is holomorphic in a domain D_2, and that D_0 is the intersection of D_1 and D_2, $D_0 = D_1 \cap D_2$ (Fig. 9.2). If $f_1(z) = f_2(z)$ at any arbitrary point z in D_0, the natural continuation of f_1 to D_2 is f_2. $f_2(z)$ is called the analytic continuation of $f_1(z)$ to D_2.

When the union of D_1 and D_2 ($D_1 \cup D_2$) is a simply-connected domain, and the analytic continuation f_2 of f_1 to D_2 is possible, it is unique. This fact can be explained as follows: If the analytic continuation does not remain unique, and $f_3(z)$ is another analytic continuation of $f_1(z)$ to D_2, then it should be $f_2(z) = f_3(z)$ in D_0, and $f_2(z) \neq f_3(z)$ in $D_2 - D_0$. This contradicts the identity theorem.

9.2.1.2 *Weierstrass' analytic continuation*

Another concrete method for analytic continuation is that produced by Weierstrass. It performs the following operation on a Taylor series.

Let the function $P(z : a)$

$$P(z : a) = \sum_{n=0}^{\infty} a_n(z - a)^n \qquad (9.7)$$

have the convergence radius R_a, which is given by a Taylor series around a ($|z - a| < R_a$). Considering the point b inside the convergence circle, the function $P(z : a)$ expands around b as

$$Q(z : b) = \sum_{n=0}^{\infty} b_n(z - b)^n, \quad b_n = \frac{1}{n!}P^{(n)}(b : a). \qquad (9.8)$$

Let the convergence radius of the series $Q(z : b)$ be R_b ($|z - b| < R_b$). At the intersection of the two circles,

$$P(z : a) = Q(z : b) \qquad (9.9)$$

so $Q(z : b)$ is the analytic continuation of $P(z : a)$ to $|z - b| < R_b$. The above procedure can be used specifically to perform analytic continuations in order to extend the holomorphic range of functions.

Example 9.3.

$$P(z : 0) = 1 + z + z^2 + \cdots \qquad (|z| < 1). \qquad (9.10)$$

Let us perform an analytic continuation of $P(z : 0)$ to the outside of the convergence circle $|z| < 1$. Select an appropriate point $z = b$ ($|b| < 1$). Performing a Taylor expansion of $P(z : 0)$ around $z = b$ and writing it as $Q(z : b)$ produces

$$b_n = \frac{1}{n!}P^{(n)}(b : 0) = \frac{1}{n!}\sum_{l \geq n} \frac{l!}{(l - n)!}b^{l-n} = \frac{1}{(1 - b)^{n+1}} \qquad (9.11a)$$

$$Q(z : b) = \sum_{n=0}^{\infty} b_n(z - b)^n = \sum_{n=0}^{\infty} \frac{(z - b)^n}{(1 - b)^{n+1}}$$

$$= \frac{1}{1 - b}\sum_{n=0}^{\infty} \left(\frac{z - b}{1 - b}\right)^n. \qquad (9.11b)$$

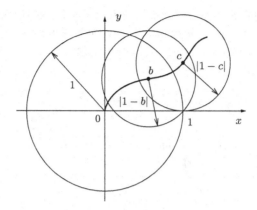

Fig. 9.3   Analytic continuation of $P(z:0) = 1 + z + z^2 + \cdots$.

The convergence radius of $Q(z:b)$ is $|1 - b|$, which is to say, it is the radius
of a circle that passes through $z = 1$ centered at b (Fig. 9.3). In this way,
an analytic continuation of $P(z:0)$ to the outside of the circle $|z| = 1$ is
possible, while avoiding a point $z = 1$. In fact, the series sum (9.10) is

$$P(z:0) = \frac{1}{1-z} \tag{9.12}$$

and $P(z:0)$ is originally a function with a pole at $z = 1$. Therefore, the
analytic continuation is performed while avoiding the pole $z = 1$. The series
sum of $Q(z:b)$ can be taken, and it is certainly Eq. (9.12).

9.2.2   Monodromy theorem

The theorem of identity tells that when two analytic continuations are
performed on one holomorphic function defined in the neighborhood of an
initial point of one fixed path, they produce the same holomorphic function.

Then, in a simply-connected domain, when analytic continuation is performed along two (or any arbitrary number of) different paths having the
common initial and final points, what happens at the final point for functions that are identical in the neighborhood of the initial point? The definition of a simply-connected domain allows continuous transformations
between two paths with fixed initial and final points. And the uniqueness
of the function is guaranteed for the analytic continuation during succes-
sive transformation of paths. Therefore, when analytic continuation is performed along different paths for functions[1] that match in the neighborhood

---

[1]Functions for which $f_1 = f_2$ in the neighborhood of a given point are called equivalent

of the initial point, uniqueness is assured at the final point. This can be summarized in the following monodromy theorem.[2]

Theorem 9.5 (Monodromy theorem). *Let us assume that a holomorphic function $f(z, a)$ is defined in the neighborhood of a point a in a simply-connected domain D, and the analytic continuation of $f(z, a)$ is possible along any arbitrary path with an initial point a in D. Then $f(z; a)$ is uniquely extended to a single-valued holomorphic function $f(z)$ in D.*

9.2.3 *Riemann surfaces*

Suppose that the holomorphic domain D_1 of $f_1(z)$ and the holomorphic domain D_2 of f_2 have two intersection D_0 and D_0', which are separate from each other (Fig. 9.4). Then, the union of D_1 and D_2 is multiply-connected. In that case, even if $f_1(z) = f_2(z)$ in D_0, we do not know whether $f_1(z) = f_2(z)$ in D_0', because nothing has been said about $f_1(z)$, $f_2(z)$ in D_0'. Let us define $f(z) = f_1(z)$, $z \in D_1$ and $f(z) = f_2(z)$, $\in D_2$. Then, when $f_1(z) \neq f_2(z)$ in D_0', $f(z)$ becomes a multi-valued function in the union of D_1 and D_2. In this case, we should consider D_1 and D_2 separately in D_0', and only D_0 connects them. This is Riemann surface in a multiply-connected domain.

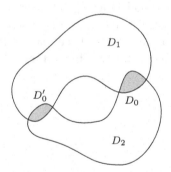

Fig. 9.4 Analytic continuation of multiply-connected domains.

Example 9.4. An exponential function is defined on the real axis

$$e^x = 1 + \frac{1}{1!}x + \frac{1}{2!}x^2 + \cdots . \qquad (9.13)$$

---

functions. A group of these equivalent functions (an equivalence class) is called a **germ**, and all the germs are called a **sheaf**.

[2]Monodromy means "running round singly (singularity)".

The analytic continuation of the above is

$$e^z = 1 + \frac{1}{1!}z + \frac{1}{2!}z^2 + \cdots. \tag{9.14}$$

In this example, the analytic continuation of the differentiable real number function $f(x)$ to the complex plane can be obtained simply by substituting z for x. If the function $f(z)$ that is obtained is single-valued holomorphic, the analytic continuation of this method is uniquely determined. When $f(z)$ is a multi-valued function, it is necessary to determine an argument so that it matches $f(x)$ in the first region. This is the most general and concrete method of analytic continuation.

9.2.4   *Natural boundary*

Consider

$$f(z) = \sum_{n=1}^{\infty} z^{n!} = z + z^2 + z^6 + z^{24} + \cdots \qquad (|z| < 1). \tag{9.15}$$

This diverges at $z = 1$, and also diverges at all points z that satisfy $z^{n!} = 1$ ($n = 1, 2, 3, \cdots$). Such points exist densely on the unit circle centered at the origin. (This means that one can find points z on a unit circle satisfying $z^{m!} = 1$ in any neighborhood of any point z' which satisfies $(z')^{n!} = 1$.) Therefore, whatever kind of point we would choose inside the unit circle, analytic continuation to the outside of the unit circle is not possible. This kind of boundary is called a **natural boundary**.

Chapter 10

Meromorphic Functions

In this chapter, we add supplementary material that is necessary when using complex function theory, or which often appears in its application.

We start by explaining the partial fraction expansion of a meromorphic function, and the infinite product representation of an entire function. This content has already been partially covered in Sec. 3.6, but we add some to the explanation in this chapter. We also cover Γ (gamma) functions and B (beta) functions. These functions often appear in physics and engineering, and they are also good examples of integration representation and analytic continuation of special functions, and of the saddle point method and asymptotic expansion.

10.1 Partial fraction expansion of a meromorphic function, and the infinite product representation of an entire function

10.1.1 *Partial fraction expansion (decomposition) of a meromorphic function*

We assume that $g(z)$ has first-order poles $b_1, b_2, \cdots, b_n, \cdots$ in a finite domain, and that no other singularities ($b_j \neq 0$), which is to say, it is a **meromorphic function** that only has first-order poles (Definition 8.9). Let the contour path C_R be a circle of radius R, centered at the origin, with all poles interior to C_R. Taking z interior to C_R,

$$\frac{1}{2\pi \mathrm{i}} \oint_{C_R} \frac{g(\zeta)}{\zeta - z} \mathrm{d}\zeta = g(z) + \sum_j \frac{B_j}{b_j - z}. \tag{10.1}$$

B_j is the residue of $g(z)$ at b_j. Substituting

$$\frac{1}{\zeta - z} = \frac{1}{\zeta} + \frac{z}{\zeta(\zeta - z)} \tag{10.2}$$

159

into the left-hand side of Eq. (10.1), we get

$$\frac{1}{2\pi i} \oint_{C_R} \frac{g(\zeta)}{\zeta - z} \, d\zeta = \frac{1}{2\pi i} \oint_{C_R} \frac{g(\zeta)}{\zeta} \, d\zeta + \frac{z}{2\pi i} \oint_{C_R} \frac{g(\zeta)}{\zeta(\zeta - z)} \, d\zeta. \tag{10.3}$$

Letting $z = 0$ in Eq. (10.1) produces

$$\frac{1}{2\pi i} \oint_{C_R} \frac{g(\zeta)}{\zeta} \, d\zeta = g(0) + \sum_j \frac{B_j}{b_j}. \tag{10.4}$$

Also, as $g(z)$ has no singularities on the z plane except first-order poles at a finite distance, *i.e.* a point of infinity is a regular point for $g(z)$,

$$\lim_{R \to \infty} \frac{1}{2\pi i} \oint_{C_R} \frac{g(\zeta)}{\zeta(\zeta - z)} \, d\zeta = 0. \tag{10.5}$$

Substituting Eqs. (10.4)–(10.5) in Eq. (10.3) and, putting it into Eq. (10.1),

$$g(z) = g(0) + \sum_j B_j \left(\frac{1}{z - b_j} + \frac{1}{b_j} \right) \tag{10.6}$$

is obtained. Conversely, if $g(z)$ is written in the form of the above equation and is holomorphic in the neighborhood of any point other than b_j, $g(z)$ is a meromorphic function for which $z = b_j$ is a first-order pole.

Example 10.1.

$$g(z) = \cot z - \frac{1}{z} \tag{10.7}$$

has first-order poles $z_k = k\pi$ $(k = \pm 1, \pm 2, \pm 3, \cdots)$ and all residues are 1.

$$B_k = \frac{\cos z}{d \sin z / dz} \bigg|_{z = z_k} = 1. \tag{10.8}$$

Also, $g(0) = 0$, so

$$\cot z = \frac{1}{z} + \sum_{k = -\infty(\neq 0)}^{\infty} \left(\frac{1}{z - k\pi} + \frac{1}{k\pi} \right) = \frac{1}{z} + \sum_{k=1}^{\infty} \frac{2z}{z^2 - k^2 \pi^2}. \tag{10.9}$$

10.1.2   *Infinite product representation of entire function*

We introduced infinite products in Sec. 3.6. The examples of $\sin z$ and $\cos z$ were also presented there, but the discussion was not completed. We will continue the discussion here in a little more detail, including its analyticity.

When there is a sequence $\{u_n\}_{n=1}^{\infty}$,

$$\prod_{n=1}^{\infty} (1 + u_n) = (1 + u_1)(1 + u_2)(1 + u_3) \cdots \tag{10.10}$$

is called an infinite product.

Theorem 10.1. *Assume that $f(z)$ is holomorphic in the finite domain of z (i.e. an entire function, see Sec. 4.3.1) and has an infinite number of first-order zero points $a_1, a_2, \cdots$ with*

$$\lim_{n \to \infty} |a_n| = \infty. \tag{10.11}$$

In that case, $f(z)$ can be expressed, using the infinite product, as

$$f(z) = f(0) \, e^{\{f'(0)/f(0)\}z} \prod_{n=1}^{\infty} \left[\left(1 - \frac{z}{a_n} \right) e^{z/a_n} \right]. \tag{10.12}$$

Conversely, if the right-hand side of Eq. (10.12) uniformly converges in the neighborhood of any point, $f(z)$ is a holomorphic function (entire function) having first-order zero points $a_1, a_2, \cdots, a_n, \cdots$.

Proof. Consider $f'(z)/f(z)$ for the entire function $f(z)$, based on the result of partial fraction expansion of a meromorphic function. $a_1, \cdots, a_n, \cdots$ are first-order zero points of $f(z)$, then these points are the first order poles of $f'(z)/f(z)$ and the residues at these points are 1. Therefore, using Eq. (10.6),

$$\frac{f'(z)}{f(z)} = \frac{f'(0)}{f(0)} + \sum_{n=1}^{\infty} \left(\frac{1}{z - a_n} + \frac{1}{a_n} \right). \tag{10.13}$$

Integrating this produces

$$\log f(z) = \frac{f'(0)}{f(0)} z + \sum_{n=1}^{\infty} \left[\log \left(1 - \frac{z}{a_n} \right) + \frac{z}{a_n} \right] + (\text{constant}). \tag{10.14}$$

Then

$$f(z) = A e^{(f'(0)/f(0))z} \prod_{n=1}^{\infty} \left[\left(1 - \frac{z}{a_n} \right) e^{z/a_n} \right]. \tag{10.15}$$

Letting $z = 0$, we get $f(0) = A$. Therefore, we obtain the infinite product representation of an entire function $f(z)$

$$f(z) = f(0) \, e^{(f'(0)/f(0))z} \prod_{n=1}^{\infty} \left[\left(1 - \frac{z}{a_n} \right) e^{z/a_n} \right]. \tag{10.16}$$

$\square$

Example 10.2.

$$f(z) = \frac{\sin z}{z} \tag{10.17}$$

is an entire function with $z_k = k\pi$ ($k = \pm 1, \pm 2, \cdots$) as first-order zero points. $f(0) = 1$, $f'(0) = 0$, then

$$\frac{\sin z}{z} = \prod_{k=1}^{\infty} \left(1 - \frac{z^2}{k^2 \pi^2} \right). \tag{10.18}$$

10.2 Γ functions and B functions

10.2.1 Γ functions and their analytic continuation

Consider a following function of a complex variable z:

$$\Gamma(z) = \int_0^\infty e^{-t} t^{z-1} dt. \tag{10.19}$$

This is called (Euler's) Γ (gamma) function, and the integration of the right-hand side is called **Euler integral of the second kind**.

If z is a positive real number $z = x > 0$, the partial integration is performed as

$$\Gamma(x+1) = \int_0^\infty e^{-t} t^x dt = -e^{-t} t^x \Big|_{t=0}^\infty + x \int_0^\infty e^{-t} t^{x-1} dt = x\Gamma(x). \tag{10.20}$$

Also, if x is 0 or a positive integer n,

$$\Gamma(1) = \int_0^\infty e^{-t} dt = 1 \tag{10.21}$$

$$\Gamma(n+1) = \int_0^\infty e^{-t} t^n dt = n!, \tag{10.22}$$

where let $0! = 1$.

If z is a complex number, and $0 < \varepsilon \leq \operatorname{Re} z \leq M < \infty$ for a certain finite number M, then

$$|e^{-t} t^{z-1}| \leq e^{-t} t^{\varepsilon-1} \quad \text{when } 0 < t \leq 1 \tag{10.23a}$$

$$|e^{-t} t^{z-1}| \leq e^{-t} t^{M-1} \quad \text{when } 1 \leq t < \infty. \tag{10.23b}$$

The integration of these evaluation formulae

$$\int_0^1 e^{-t} t^{\varepsilon-1} dt, \qquad \int_1^\infty e^{-t} t^{M-1} dt$$

are both finite and determinate, so the integration

$$\int_0^\infty e^{-t} t^{z-1} dt \tag{10.24}$$

converges for z in $\varepsilon \leq \operatorname{Re} z \leq M$. As ε, M are arbitrary, the above integration (10.19) converges to a finite value at $\operatorname{Re} z > 0$. Therefore, $\Gamma(z)$ is continuous over $\operatorname{Re} z > 0$. Next, suppose that C is a closed Jordan curve in the domain $\operatorname{Re} z > 0$, then the uniform convergence allows the integral

$$\oint_C \Gamma(z) \, dz = \int_0^\infty e^{-t} dt \oint_C t^{z-1} dz. \tag{10.25}$$

From Cauchy's theorem, $\oint_C t^{z-1}\mathrm{d}z = 0$, so

$$\oint_C \Gamma(z)\,\mathrm{d}z = 0. \tag{10.26}$$

Which is to say, $\Gamma(z)$ is holomorphic in $\operatorname{Re} z > 0$. Therefore, we can perform analytic continuation of Eq. (10.20) to the region $\operatorname{Re} z > 0$, and obtain

$$\Gamma(z+1) = z\Gamma(z). \tag{10.27}$$

If this is rewritten as

$$\Gamma(z) = \frac{\Gamma(z+1)}{z}, \tag{10.28}$$

the right-hand side is meromorphic in $\operatorname{Re}(z+1) > 0$ and has $z = 0$ as a first-order pole. This means that analytic continuation is performed on $\Gamma(z)$ in $0 > \operatorname{Re} z > -1$. Furthermore, if a Γ function extended to $0 > \operatorname{Re} z > -1$ is used for Eq. (10.28), the right hand side undergoes analytic continuation to $-1 > \operatorname{Re} z > -2$, and has a first-order pole $z = -1$ as

$$\Gamma(z) = \frac{\Gamma(z+2)}{z(z+1)}.$$

This can be repeated to extend $\Gamma(z)$ onto the whole z plane, and obtain

$$\Gamma(z) = \frac{\Gamma(z+3)}{z(z+1)(z+2)} = \cdots = \frac{\Gamma(z+n+1)}{z(z+1)\cdots(z+n)}. \tag{10.29}$$

Therefore, $z = 0, -1, -2, \cdots -n, \cdots$ are first-order poles. From the equation above, the residue at $z = -n$ can be seen to be

$$\lim_{z \to -n} (z+n)\Gamma(z) = \frac{1}{(-n)(-n+1)\cdots(-1)} = \frac{(-1)^n}{n!}. \tag{10.30}$$

Figure 10.1 shows the behavior of Γ function on the real axis.

The equation (10.19) was defined in the region $\operatorname{Re} z > 0$. Let us modify this and derive an integral representation of Γ function that is valid for all complex numbers z. Consider the integration along the contour path in Fig. 10.2(a);

$$I(z) = \int_C e^{-\zeta}\zeta^{z-1}\mathrm{d}\zeta. \tag{10.31}$$

This integration requires the contour path C not to pass through the origin. On the contour path C, taking $\zeta = \rho\,e^{i\phi}$, ζ^{z-1} is

$$\zeta^{z-1} = e^{(z-1)(\ln \rho + i\phi)} \qquad (0 \le \phi \le 2\pi). \tag{10.32}$$

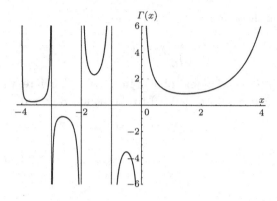

Fig. 10.1   Behavior of $\Gamma(x)$.

Moving along the upper side of the real axis on the ζ plane, we choose the argument of ζ to be $\phi = 0$. Here, let us continuously transform the contour path C to the contour path in Fig. 10.2(b);

$$I(z) = \int_{\infty}^{a} + \int_{abc} + \int_{c}^{\infty} . \tag{10.33}$$

When in the first integration of Eq. (10.33), the point a (coordinates $(r, 0)$) approaches the origin O, it produces

$$\int_{\infty}^{a} = \int_{\infty}^{r} e^{-\rho}\rho^{z-1}d\rho \to -\Gamma(z) \qquad (r \to 0). \tag{10.34}$$

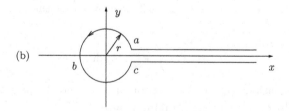

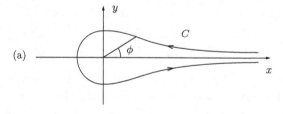

Fig. 10.2   Path of integration C of the Γ Function.

As the second integration in Eq. (10.33) $\zeta = r\,e^{i\phi}$, $d\zeta = ir\,e^{i\phi}d\phi$, so

$$\int_{abc} = \int_0^{2\pi} ir\,e^{(z-1)\ln r}e^{-re^{i\phi}}e^{i(z-1)\phi}e^{i\phi}d\phi$$

$$= ir\,e^{(z-1)\ln r}\int_0^{2\pi} d\phi\,e^{-re^{i\phi}}e^{iz\phi}$$

$$= ir^z\int_0^{2\pi} d\phi\,e^{-re^{i\phi}}e^{iz\phi}. \tag{10.35}$$

As long as $\operatorname{Re} z > 0$, $\int_{abc} \to 0$ when $r \to 0$.

On the contour path of the third integration in Eq. (10.33) the argument of ζ is $\phi = 2\pi$, so ζ^{z-1} is

$$\zeta^{z-1} = e^{(z-1)(\ln \rho + 2\pi i)} = \rho^{z-1}e^{2\pi i(z-1)}. \tag{10.36}$$

Therefore, the third integration is

$$\int_c^\infty = \int_r^\infty d\rho\,e^{2\pi i}e^{-\rho}\rho^{z-1}e^{2\pi i(z-1)}$$

$$= e^{2\pi iz}\int_r^\infty d\rho\,e^{-\rho}\rho^{z-1} \to e^{2\pi iz}\Gamma(z) \qquad (r \to 0). \tag{10.37}$$

From the above, it follows that, when $\operatorname{Re} z > 0$,

$$I(z) = (e^{2\pi iz} - 1)\,\Gamma(z) = 2ie^{\pi iz}\sin \pi z\Gamma(z) \tag{10.38}$$

is true. Thus, we see that $I(z)/(2ie^{\pi iz}\sin \pi z)$ is defined for the whole area of z, and is equal to $\Gamma(z)$ for $\operatorname{Re} z > 0$. Therefore, $I(z)/(2i\,e^{\pi iz}\sin \pi z)$ is the analytic continuation of $\Gamma(z)$ to the whole area of z;

$$\Gamma(z) = \frac{e^{-\pi iz}}{2i\sin \pi z}\int_C e^{-\zeta}\zeta^{z-1}d\zeta. \tag{10.39}$$

It is noticed that this equation does not hold when $z = 0, 1, 2, \cdots$.

10.2.2 B functions

The following is a function closely related to Γ functions:

$$B(p,q) = \int_0^\infty \frac{x^{q-1}}{(1+x)^{p+q}}dx \qquad (\operatorname{Re} p, \quad \operatorname{Re} q > 0). \tag{10.40}$$

This is a B (beta) function, and the integral form is called **Euler integral of the first kind**. Letting $x = (1-t)/t$,

$$B(p,q) = \int_0^1 t^{p-1}(1-t)^{q-1}dt. \tag{10.41}$$

Also, letting $t = 1 - s$ the roles of p and q gives are interchanged, then
$$B(p,q) = B(q,p). \tag{10.42}$$
Letting $t \to kt$ $(k > 0)$ in the Euler integration of the second kind obtains
$$\frac{\Gamma(z)}{k^z} = \int_0^\infty e^{-kt} t^{z-1} \mathrm{d}t. \tag{10.43}$$
Letting $k = 1 + s$, $z = p + q$ here,
$$\frac{\Gamma(p+q)}{(1+s)^{p+q}} = \int_0^\infty e^{-(1+s)t} t^{p+q-1} \mathrm{d}t.$$
Multiplying both sides by s^{p-1} and integrating for s between 0 and ∞ produces
$$\Gamma(p+q) \int_0^\infty \frac{s^{p-1}}{(1+s)^{p+q}} \mathrm{d}s = \int_0^\infty \mathrm{d}s \int_0^\infty \mathrm{d}t \, e^{-(1+s)t} t^{p+q-1} s^{p-1}.$$
It can be rewritten as
$$\text{(Right-hand side)} = \int_0^\infty e^{-t} t^{q-1} \mathrm{d}t \int_0^\infty e^{-st} (st)^{p-1} t \, \mathrm{d}s = \Gamma(q)\,\Gamma(p)$$
$$\text{(Left-hand side)} = \Gamma(p+q)\,B(p,q).$$
Thus,
$$B(p,q) = \frac{\Gamma(p)\,\Gamma(q)}{\Gamma(p+q)}. \tag{10.44}$$
In the above, we assumed that $\operatorname{Re} p$, $\operatorname{Re} q > 0$, but as the right-hand side of Eq. (10.44) is defined for the whole area of p, q, $B(p, q)$ is also determined by this equation for the whole area of p, q.

In particular, letting $p = 1 - z$, $q = z$, $1 > \operatorname{Re} z > 0$ in Eq. (10.40) and using the Eq. (6.70) in Example 6.18, we get
$$B(1-z, z) = \int_0^\infty \frac{x^{z-1}}{1+x} \mathrm{d}x = \frac{\pi}{\sin \pi z}. \tag{10.45}$$
Paying attention to $\Gamma(1) = 1$ produces
$$\Gamma(1-z)\Gamma(z) = \frac{\pi}{\sin \pi z}. \tag{10.46}$$
Equation (10.46) was found for $1 > \operatorname{Re} z > 0$, but with analytic continuation, it holds for all values of z. Also, from the equation above, it is clear that $\Gamma(z)$ has no zero points. Furthermore, letting $z = 1/2$ in Eq. (10.46) immediately produces
$$\Gamma\left(\frac{1}{2}\right) = \sqrt{\pi} \tag{10.47}$$
and the recurrence formula (10.27) can be used to derive
$$\Gamma\left(\frac{3}{2}\right) = \frac{1}{2}\sqrt{\pi}, \quad \Gamma\left(\frac{5}{2}\right) = \frac{1\cdot 3}{2^2}\sqrt{\pi}, \quad \Gamma\left(\frac{7}{2}\right) = \frac{1\cdot 3\cdot 5}{2^3}\sqrt{\pi}, \quad \cdots. \tag{10.48}$$

10.2.3 *Stirling's formula and asymptotic expansion of Γ function*

10.2.3.1 *Stirling's formula*

When n is a large integer, the value of $n!$ can be approximated to

$$n! \simeq \sqrt{2\pi n}\, n^n e^{-n}. \tag{10.49}$$

This is called Stirling's formula. Let us derive it.

In Eq. (10.19)

$$\Gamma(x+1) = \int_0^\infty e^{-t} t^x dt,$$

transformation from t to τ taking $t = x\tau$ $(x > 0)$ produces

$$\Gamma(x+1) = x^{x+1} \int_0^\infty e^{-x\tau} \tau^x d\tau$$

$$= x^{x+1} e^{-x} \int_0^\infty e^{-x(\tau - 1 - \ln\tau)} d\tau. \tag{10.50}$$

Letting $f(\tau) = \tau - 1 - \ln\tau$ produces

$$\Gamma(x+1) = x^{x+1} e^{-x} I(x),$$

$$I(x) = \int_0^\infty e^{-xf(\tau)} d\tau. \tag{10.51}$$

Looking closely at this integration (10.51) shows that the integration in $I(x)$ does not give the same contribution throughout the whole domain of $0 < \tau < \infty$.

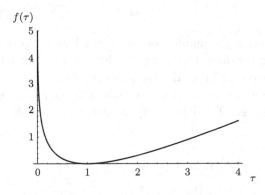

Fig. 10.3 $f(\tau) = \tau - 1 - \ln\tau$.

$f(\tau)$ varies as shown in Fig. 10.3. When $x \gg 1$, $e^{-xf(\tau)}$ reaches its maximum value at $\tau = 1$, and drops rapidly to 0 with distance from $\tau = 1$. Therefore, it is good enough to only calculate the contribution accurately from the neighborhood of $\tau = 1$. This kind of assessment method is called the **saddle point method**. Expand $f(\tau)$ around $\tau = 1$ to produce

$$f(\tau) = \frac{(\tau - 1)^2}{2} - \frac{(\tau - 1)^3}{3} + \frac{(\tau - 1)^4}{4} - \cdots. \qquad (10.52)$$

From this, it follows that

$$I(x) \approx \int_{1-\varepsilon}^{1+\varepsilon} \exp\left[-\frac{x}{2}(\tau - 1)^2\right] d\tau$$

$$= \sqrt{\frac{2}{x}} \int_{-\varepsilon\sqrt{x/2}}^{\varepsilon\sqrt{x/2}} e^{-u^2} du \approx \sqrt{\frac{2}{x}} \int_{-\infty}^{\infty} e^{-u^2} du = \sqrt{\frac{2\pi}{x}}. \qquad (10.53)$$

ε here is a small positive number, but x was sufficiently large that it is reasonable to take $\varepsilon\sqrt{x/2} \to \infty$. Therefore, from Eqs. (10.51) and (10.53), we obtain

$$\Gamma(x + 1) \sim \sqrt{2\pi x}\, x^x e^{-x} \qquad (x \gg 1). \qquad (10.54)$$

Letting $x = n$ (positive integer) produces

$$n! \sim \sqrt{2\pi n}\, n^n e^{-n}. \qquad (10.55)$$

This is Stirling's formula.

If the expansion of $f(\tau)$ is used to a high order, the high-order expansion for $\Gamma(x)$ where $x \gg 1$ produces the following:

$$\Gamma(x) \sim \sqrt{2\pi} x^{x-1/2} e^{-x}\left(1 + \frac{1}{12x} + \frac{1}{288x^2} - \frac{139}{51840x^3} - \frac{571}{2488320x^4} + \cdots\right). \qquad (10.56)$$

Initially, increasing the number of terms will better represent the value. However, there is a limit to this approach, and if the number of terms is increased beyond that limit, the value deviates from the correct one. Furthermore, the series sum calculated to an infinite number of terms does not necessarily converge. This kind of expansion equation is called **asymptotic expansion**.

10.2.3.2　*Asymptotic expansion*

We will now summarize the general theory of asymptotic expansion and asymptotic series.

Definition 10.1. For the function $f(z)$ and the finite sum $s_n(z)$

$$s_n(z) = a_0 + \frac{a_1}{z} + \frac{a_2}{z^2} + \cdots + \frac{a_n}{z^n}, \tag{10.57}$$

let us define

$$z^n R_n(z) = z^n[f(z) - s_n(z)]. \tag{10.58}$$

Here, in the following cases

$$\text{if } n \text{ is fixed } \lim_{|z| \to \infty} z^n R_n(z) = 0 \tag{10.59}$$

$$\text{if } z \text{ is fixed } \lim_{n \to \infty} z^n R_n(z) = \infty, \tag{10.60}$$

$S(z) \equiv \lim_{n \to \infty} s_n$ is called asymptotic expansion. It is written

$$f(z) \sim \sum_{n=0}^{\infty} a_n z^{-n}. \tag{10.61}$$

The series on the right-hand side is called **asymptotic series**. In that case, $S(z)$ may or may not converge.

Chapter 11

Elliptic Integrals and Elliptic Functions

Elliptic functions are the central achievement of 19th-century classical analysis. They include much important content, and appear in many problems, such as the motion of a spinning top, the motion of celestial bodies, various questions in fields such as analytical mechanics, and flat-plate aerofoils in a flow bounded by walls. An (incomplete) elliptic integral is the inverse function of an elliptic function, and its name comes from the problem of the arc length of an ellipse.

11.1 Elliptic Integrals

11.1.1 *Definition of elliptic integral*

We will start with a few examples to show what, specifically, an elliptic integral is, and what kinds of problem it appears in.

Definition 11.1. Assume that $R(z, w)$ is a rational function of z and w, $g(z)$ is a third- or fourth-order polynomial and $g(z) = 0$ does not have multiple roots. The integral of the form of

$$\int R(z, \sqrt{g(z)})\, \mathrm{d}z \tag{11.1}$$

is called an **elliptic integral**.

This indefinite integral cannot be an elementary function.

171

11.1.2   *Standard forms of elliptic integral*

Equation (11.1) is generally known to take the following three forms:

$$\int_0^z \frac{dz}{\sqrt{(1-z^2)(1-k^2z^2)}} \qquad \text{(Elliptic integral of the first kind)}$$

$$(11.2)$$

$$\int_0^z \sqrt{\frac{1-k^2z^2}{1-z^2}}\,dz \qquad \text{(Elliptic integral of the second kind)}$$

$$(11.3)$$

$$\int_0^z \frac{dz}{(1+nz^2)\sqrt{(1-z^2)(1-k^2z^2)}} \qquad \text{(Elliptic integral of the third kind)}$$

$$(11.4)$$

These equations are called **complete elliptic integrals** when the upper limit of integral is 1, and **incomplete elliptic integrals** when the upper limit of integral is not 1. Integrals Eqs. (11.2)–(11.4) were introduced by A. M. Legendre, and are called **Legendre-Jacobi standard forms**. Each is named as in (11.2)–(11.4). If these are regarded as the integrals of multi-valued functions, they are Schwarz–Christoffel transformation discussed in Example 6.19.

11.1.3   *Example of elliptic integral*

Example 11.1. Let us show that the arc length of an ellipse is written as an elliptic integral of the second kind. The length of an arc of an ellipse

$$\frac{x^2}{a^2} + \frac{y^2}{b^2} = 1 \qquad (0 < b < a) \tag{11.5}$$

is

$$s = \int \sqrt{(dx)^2 + (dy)^2}. \tag{11.6}$$

When we perform a transformation, taking $x = a\sin\theta$, $y = b\cos\theta$,

$$s = a\int_0^\phi \sqrt{1 - k^2\sin^2\theta}\,d\theta, \quad \text{where} \quad k = \sqrt{\frac{a^2-b^2}{a^2}}. \tag{11.7}$$

Furthermore, taking $z = \sin\theta$,

$$s = a\int_0^{\sin\phi} \sqrt{\frac{1-k^2z^2}{1-z^2}}\,dz. \tag{11.8}$$

This is the elliptic integral of the second type.

Similarly, the arc length of a cycloid $x = r(\theta - \sin\theta)$, $y = r(1 - \cos\theta)$ is written as an elliptic integral of the second kind (Fig. 11.1(a)). The arc length of lemniscate $r^2 = a^2 \cos 2\theta$ is expressed as an elliptic integral of the first kind (Fig. 11.1(b)). Beyond such curve arc length problems, elliptic integrals are used for various problems in physics, buckling problems, the external shape of cylindrical water bags, and the shapes formed by surface tension in a water surface between two parallel plates ([M. Toda (2001)]).

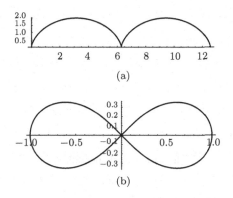

Fig. 11.1   (a) cycloid $x = (\theta - \sin\theta)$, $y = (1 - \cos\theta)$ and (b) lemniscate $r^2 = \cos 2\theta$.

Example 11.2. Consider the problem of a single pendulum.   As in Fig. 11.2, let the angle be ϕ, and let the speed of a particle (of mass m) attached to one end of the pendulum be $v_0 = l\omega_0$ at $\phi = 0$ (where l is the length of a light rod joining the particle to the fixed point O, and ω_0 is the angular velocity of rotation), where gravitational acceleration is g and time is t. The equation of motion for this system is

$$\frac{\mathrm{d}^2\phi}{\mathrm{d}t^2} = -\frac{g}{l}\sin\phi. \tag{11.9}$$

The pendulum oscillates when $v_0^2 < 4gl$. Considering this situation only, and letting α be the maximum angle of oscillation, α is

$$\sin^2\frac{\alpha}{2} = \frac{v_0^2}{4gl}. \tag{11.10}$$

Furthermore, when we define

$$k = \sin\frac{\alpha}{2}, \qquad \sin\frac{\phi}{2} = kz, \tag{11.11}$$

the period of oscillation is given by

$$T = 4\sqrt{\frac{l}{g}} \int_0^1 \frac{\mathrm{d}z}{\sqrt{(1 - z^2)(1 - k^2 z^2)}}. \tag{11.12}$$

This is a complete elliptic integral of the first kind.

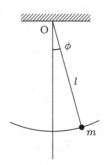

Fig. 11.2   Single pendulum.

11.1.4   *Properties of elliptic integral*

Let us consider the properties of an elliptic integral of the first kind where $0 < k < 1$;

$$w(z) = \int_0^z \frac{\mathrm{d}z}{\sqrt{(1 - z^2)(1 - k^2 z^2)}}. \tag{11.13}$$

The integrated function

$$f(z) = \frac{1}{\sqrt{(1 - z^2)(1 - k^2 z^2)}} \tag{11.14}$$

is a multi-valued function with branch points at

$$z = 1, \quad -1, \quad \frac{1}{k}, \quad -\frac{1}{k}. \tag{11.15}$$

Assume a path of integration as shown in Fig. 11.3(a) (avoiding branch points in a clockwise direction). Let us consider what kind of domain the upper half of the z plane maps to (Schwarz–Christoffel transformation Example 6.19). Specify that $f(z) > 0$ (the argument of $f(z)$ is 0) in $-1 < z < 1$. In that case, the argument of $f(z)$ at $z = 1$ increases by just $+\pi/2$, and by a further $+\pi/2$ at $z = 1/k$. Conversely, if z moves to the left from

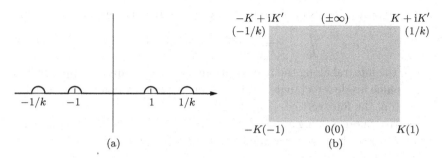

Fig. 11.3 (a) To calculate an elliptic integral of the second kind, branch points are avoided in the upper half of the plane. (b) A domain mapped from the upper half of the z plane. Corresponding points on the real axis in the z plane are shown in parentheses.

the origin, the argument of $f(z)$ at $z = -1$ decreases by just $+\pi/2$, and by a further $+\pi/2$ at $z = -1/k$. Defined K and K' as

$$K = \int_0^1 \frac{dx}{\sqrt{(1 - x^2)(1 - k^2 x^2)}} \tag{11.16a}$$

$$K' = \int_1^{1/k} \frac{dx}{\sqrt{(x^2 - 1)(1 - k^2 x^2)}}, \tag{11.16b}$$

then points $z = \pm 1,\ \pm 1/k$ map to

$$w = \pm K, \qquad \pm K + iK' \tag{11.17}$$

by Eq. (11.13), respectively. The argument of $f(z)$, $z = x > 1/k$ on the real axis is π. Therefore, the contribution from $1/k < x < \infty$ to the integral, with the transformation $x = 1/(ky)$, can be calculated as

$$\int_{1/k}^\infty \frac{dz}{\sqrt{(1 - z^2)(1 - k^2 z^2)}} = -\int_{1/k}^\infty \frac{dx}{\sqrt{(x^2 - 1)(k^2 x^2 - 1)}}$$

$$= -\int_0^1 \frac{dy}{\sqrt{(1 - y^2)(1 - k^2 y^2)}}$$

$$= -K. \tag{11.18}$$

Therefore,

$$\int_0^\infty \frac{dz}{\sqrt{(1 - z^2)(1 - k^2 z^2)}} = \int_0^1 + \int_1^{1/k} + \int_{1/k}^\infty = K + \int_1^{1/k} + (-K)$$

$$= i \int_1^{1/k} \frac{dx}{\sqrt{(x^2 - 1)(1 - k^2 x^2)}} = iK'. \tag{11.19}$$

Similarly, the integration from 0 to $-\infty$ on the real axis gives

$$\int_0^{-\infty} \frac{dz}{\sqrt{(1-z^2)(1-k^2z^2)}} = iK'. \tag{11.20}$$

Thus, the integral along the real axis from $-\infty$ to ∞ maps the upper half of the z plane inside a rectangle on the w plane, with each point in Eq. (11.17) as each of the four vertices (Fig. 11.3(b)).

Furthermore, setting $\arg(f(z')) = 0$ at $z' = 0$, the value of direct integral from 0 to z without going around any branch point is $w(z)$:

$$w(z) = \int_0^z \frac{dz'}{\sqrt{(1-z'^2)(1-k^2z'^2)}}. \tag{11.21}$$

Then, setting the $\arg(f(z')) = 0$ at the starting point $(z' = 0)$, the value of the integral from 0 to z after going around the point $z = 1$ clockwise is $2K - w(z)$. Also, if a path is added, before the integration path from 0 ($\arg z' = 0$, $\arg f(z') = 0$) to z, that goes around the point $z' = 1$ counterclockwise, the value of the integral is $-2K + w(z)$, since the argument of $f(z')$ at the starting point $z' = 0$ of the whole integration path is $-\pi$.

11.1.5   Elliptic integral and Jacobi's elliptic functions

In general, the inverse function of $w(z)$ defined in Eq. (11.13) is written as

$$z = \mathrm{sn}\, w. \tag{11.22}$$

The function $\mathrm{sn}\, w$ is an elliptic function (Jacobi elliptic function). Let us study this function and see that it has the fundamental periods $4K$ and $2iK'$. Write a rectangular domain on the complex w plane shown in Fig. 11.3(b) as Δ.

(1) $\mathrm{sn}\, w$ is a real number on the real axis in Δ. According to the reflection principle, the function $\mathrm{sn}\, w$ is extended to a domain Δ' which is symmetric with respect to the real axis. If $w \in \Delta$, we see

$$\mathrm{sn}\,(\overline{w}) = \overline{\mathrm{sn}\,(w)}$$

at the point $\overline{w} \in \Delta'$.

As we have already seen, $\mathrm{sn}\,(w)$ is real at w on the line $\mathrm{Re}\, w = K$. Therefore, the same is true for a domain Δ'' which is a reflection image of Δ with respect to a straight line vertical to the real axis at $w = K$ and parallel to the imaginary axis. Through these processes, the function $\mathrm{sn}\,(w)$ defined in Δ has analytic continuation to a holomorphic function in the domain

$$\Delta_0 = \{-K < \mathrm{Re}\, w < 3K,\ -K' < \mathrm{Im}\, w < K'\}.$$

(2) Next, let us examine any arbitrary $w \in \Delta$. For w, when $w''(\in \Delta'')$ is a reflection point with respect to the straight line $\mathrm{Re}\, w = K$, and w_1 is a reflection point of w'' with respect to $\mathrm{Re}\, w = 3K$, then $w_1 = 4K + w$. In that case, $\mathrm{sn}\, w_1 = \overline{\mathrm{sn}\, w''} = \mathrm{sn}\, w$. Which is to say, $\mathrm{sn}\,(w + 4K) = \mathrm{sn}\,(w)$. If we use the theorem of identity, this is satisfied for any arbitrary w:

$$\mathrm{sn}\,(w + 4K) = \mathrm{sn}\,(w). \qquad (11.23\mathrm{a})$$

In the same way,

$$\mathrm{sn}\,(w + 2\mathrm{i}K') = \mathrm{sn}\,(w) \qquad (11.23\mathrm{b})$$

can also be shown.

The above demonstrates that $\mathrm{sn}\, w$ has two fundamental periods, $4K$ and $2\mathrm{i}K'$.

11.2 Elliptic Functions

11.2.1 *Double period*

11.2.1.1 *Singly periodic functions*

The function e^z satisfies $\mathrm{e}^{z+2\pi\mathrm{i}} = \mathrm{e}^z$. In this case, $2\pi\mathrm{i}$ is called the period of e^z. When a function $f(z)$, defined in a domain, satisfies

$$f(z + \omega) = f(z) \qquad (11.24)$$

for any arbitrary z in the domain, ω is called the period. If w is a period of $f(z)$, $-\omega, \pm 2\omega, \pm 3\omega, \cdots$ are also the period of $f(z)$. When all periods are expressed as integer multiples $(\pm\omega, \pm 2\omega, \cdots)$ of the period ω, $f(z)$ is called a singly periodic function, and ω is called the **fundamental period**.

11.2.1.2 *Double periodic function*

When there are two complex numbers ω_1, ω_2 which cannot be represented by integer multiples of each other, and they are periods of $f(z)$, and cannot be represented by any other period, $f(z)$ is called a double periodic function. In that situation, where $m_1\ m_2$ are 0, positive or negative integers,

$$m_1\omega_1 + m_2\omega_2 \qquad (11.25)$$

is the period of $f(z)$. When all periods can be expressed in this way, ω_1, ω_2 is called the **fundamental period**.

11.2.1.3   *Parallelogram produced by double period on Gauss plane: Fundamental period parallelogram*

The whole period of a double periodic function is

$$\Omega = \{m\omega_1 + n\omega_2 | m, n \in \mathbb{Z}\}, \tag{11.26}$$

where $\mathbb{Z}$ is a set of 0, positive and negative integers. This set forms an additive group (module).[1] A parallelogram domain with vertices at a, $a + \omega_1$, $a + \omega_1 + \omega_2$, $a + \omega_2$

$$\{a + \mu\omega_1 + \nu\omega_2 | 0 \le \mu < 1,\ 0 \le \nu < 1\} \tag{11.27}$$

is called a **fundamental period parallelogram** for Ω.

When

$$z_1 = z_2 + m\omega_1 + n\omega_2 \qquad (m, n \in \mathbb{Z}), \tag{11.28}$$

it is written as

$$z_1 = z_2 \qquad (\text{mod } \Omega). \tag{11.29}$$

11.2.2   *Definition of elliptic function*

Definition 11.2. When $f(z)$ is a double periodic function with no singular points other than poles in a finite domain of the z plane (meromorphic for $|z| < \infty$), it is called an **elliptic function**.

Comparing Eqs. (11.21), (11.22) and integral

$$y \equiv \int_0^x \frac{1}{\sqrt{1 - x^2}} dx \tag{11.30a}$$

$$x = \sin y, \tag{11.30b}$$

one can see that elliptic functions are a kind of generalization of trigonometric functions (Fig. 11.4).

Before discussing the properties of elliptic functions, we should define the order of an elliptic function.

Definition 11.3 (Order of elliptic functions). The order of an elliptic function is the total of the orders of the poles included within the fundamental period parallelogram of the function.

---

[1] Also called an Abelian group or a commutative group, it is a group for which multiplication is commutative.

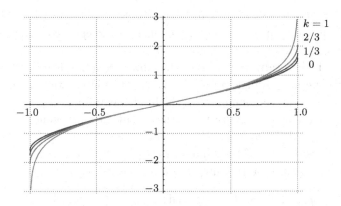

Fig. 11.4 $u(x) = \operatorname{sn}^{-1}x$ $(k = 0, 1/3, 2/3, 1)$. $u(x)$ with $k = 0$ and $k = 1$ correspond to $\sin^{-1}x$ and $\tanh^{-1}x$, respectively.

11.2.3 Basic properties of elliptic functions

A list of the general properties of elliptic functions is shown, which can be derived from their definition. Then the proofs will be followed.

Theorem 11.1 (Properties of elliptic functions). *The basic properties of elliptic functions are stated as follows:*

(a) An elliptic function, except a constant, has a pole within its fundamental period parallelogram.

(b) Sum of the residues of poles of an elliptic function is 0 in the fundamental period parallelogram.

(c) There are no elliptic functions of order 1.

(d) An elliptic function of order r (≥ 0) takes any complex value (c) r-times in the fundamental period parallelogram.

(e) If the poles of an elliptic function $f(z)$ of order r (≥ 0), are a_1, a_2, $\cdots$, a_r, and the zero points of $g(z) = f(z) - c$ are b_1, b_2, b_3, $\cdots$, b_r, then

$$a_1 + a_2 + a_3 + \cdots + a_r = b_1 + b_2 + b_3 + \cdots + b_r \qquad (\bmod \ \Omega). \quad (11.31)$$

(When the order of a pole or a zero point is k, a_l or k_l is counted k-times.)

Followings are the proofs for (a) to (e) under Theorem 11.1.

Proof. (a) [An elliptic function, except a constant, has a pole within its fundamental period parallelogram. This means that an elliptic function of order 0 is a constant.] This is true because, according to Liouville's theorem, no bounded holomorphic functions except a constant exist on the whole plane.

(b) [Sum of the residues of poles of the elliptic function is 0 in the fundamental period parallelogram.] Consider a line integration along a path C traversed on the boundary of the fundamental period parallelogram. If a pole lies on the integration path C, shift the start point a slightly to avoid the pole. Due to periodicity, the values of the elliptic function on the opposite sides of the fundamental period parallelogram are equal, and the direction of the integration is reversed, then we get

$$
\begin{aligned}
\oint_C f(z)\,\mathrm{d}z &= \int_a^{a+\omega_1} f(z)\,\mathrm{d}z + \int_{a+\omega_1}^{a+\omega_1+\omega_2} f(z)\,\mathrm{d}z \\
&\quad + \int_{a+\omega_1+\omega_2}^{a+\omega_2} f(z)\,\mathrm{d}z + \int_{a+\omega_2}^{a} f(z)\,\mathrm{d}z \\
&= \left(\int_a^{a+\omega_1} f(z)\,\mathrm{d}z + \int_{a+\omega_1+\omega_2}^{a+\omega_2} f(z)\,\mathrm{d}z \right) \\
&\quad + \left(\int_{a+\omega_1}^{a+\omega_1+\omega_2} f(z)\,\mathrm{d}z + \int_{a+\omega_2}^{a} f(z)\,\mathrm{d}z \right) \\
&= \left(\int_a^{a+\omega_1} f(z)\,\mathrm{d}z + \int_{a+\omega_1}^{a} f(z)\,\mathrm{d}z \right) \\
&\quad + \left(\int_a^{a+\omega_2} f(z)\,\mathrm{d}z + \int_{a+\omega_2}^{a} f(z)\,\mathrm{d}z \right) \\
&= 0. \tag{11.32}
\end{aligned}
$$

An elliptic function except for a constant always has poles, and the integration along a closed path is 0. So the sum of the residues is 0 in the fundamental period parallelogram.

(c) [There are no elliptic functions of order 1.] An elliptic function of order 1 would have always a single first-order pole. As a first-order pole always has a non-zero residue, this contradicts (b). Therefore, there is no elliptic function of order 1.

(d) [An elliptic function of order $r \geq 0$ takes any complex value r-times in the fundamental period parallelogram.] If $f(z)$ is an elliptic function of order r, $f(z) - c$ is also an elliptic function having the same period and the same order. Therefore, it is enough to show that there are r zero points of $f(z) - c$, including multiplicity. $g(z) = f'(z)/f(z)$ is an elliptic function having the same period as $f(z)$. According to the argument principle, when the total of the orders of zero points of $g(z)$

in a domain is N, and the total of the orders of poles is P (this is the order r of the elliptic function),

$$\frac{1}{2\pi i} \oint_C g(z) \, dz = N - P \tag{11.33}$$

is true where C is the boundary of a fundamental period parallelogram. From (b), the left-hand side of Eq. (11.33) is 0, and therefore, $N = P = r$.

(e) [If the poles of an elliptic function $f(z)$ of order $r \geq 0$ are $a_1, a_2, \cdots, a_r$, and the zero points of $g(z) = f(z) - c$ are $b_1, b_2, b_3, \cdots, b_r$, then $a_1 + a_2 + a_3 + \cdots + a_r = b_1 + b_2 + b_3 + \cdots + b_r \pmod{\Omega}$]. If a is (one of the) l-th order poles of $f(z) - c$, and b is similarly one of the m-th order zero points, then one can expand in the neighborhood of each pole or zero point as

$$z\frac{d}{dz}\log[f(z) - c] = \frac{-la}{z - a} + (\text{holomorphic part at } z = a) \tag{11.34a}$$

$$z\frac{d}{dz}\log[f(z) - c] = \frac{mb}{z - b} + (\text{holomorphic part at } z = b) \tag{11.34b}$$

respectively. Therefore, if the boundary of the fundamental period parallelogram is C,

$$\frac{1}{2\pi i} \oint_C z\frac{d}{dz}\log[f(z) - c]\,dz = \sum_{l=1}^{r} b_l - \sum_{m=1}^{r} a_m. \tag{11.35}$$

Here l and m in Eqs. (11.34a) and (11.34b) are their "multiplicity" and they are already counted in Eq. (11.35).

The boundary C of the period parallelogram is divided and named as follows ($C = C_1 + C_2 + C_3 + C_4$):

- $C_1 : a \to a + \omega_1$
- $C_2 : a + \omega_1 \to a + \omega_1 + \omega_2$
- $C_3 : a + \omega_1 + \omega_2 \to a + \omega_2$
- $C_4 : a + \omega_2 \to a$

The corresponding integrations along each boundary are

$$\int_{C_3} z\frac{d}{dz}\log[f(z) - c]\,dz = -\int_{C_1}(z + \omega_2)\frac{d}{dz}\log[f(z + \omega_2) - c]\,dz$$

$$= -\int_{C_1}(z + \omega_2)\frac{d}{dz}\log[f(z) - c]\,dz \tag{11.36a}$$

$$\int_{C_2} z\frac{d}{dz}\log[f(z) - c]\,dz = -\int_{C_4}(z + \omega_1)\frac{d}{dz}\log[f(z) - c]\,dz. \tag{11.36b}$$

Therefore,

$$\int_C z \frac{\mathrm{d}}{\mathrm{d}z} \log[f(z) - c]\, \mathrm{d}z$$

$$= -\omega_2 \int_{C1} \frac{\mathrm{d}}{\mathrm{d}z} \log[f(z) - c]\, \mathrm{d}z - \omega_1 \int_{C4} \frac{\mathrm{d}}{\mathrm{d}z} \log[f(z) - c]\, \mathrm{d}z$$

$$= -\omega_2 \Big[\log[f(z) - c] \Big]_a^{a+\omega_1} - \omega_1 \Big[\log[f(z) - c] \Big]_{a+\omega_2}^a. \qquad (11.36c)$$

The last line of Eq. (11.36c) can be rearranged to $2\pi\mathrm{i} \times$ (integer), paying attention to $|\log f(a) - c| = |\log f(a + \omega_1) - c|$ etc. Therefore,

$$\frac{1}{2\pi\mathrm{i}} \int_C z \frac{\mathrm{d}}{\mathrm{d}z} \log[f(z) - c] = m\omega_1 + n\omega_2. \qquad (11.37)$$

Here, m, n are either 0, positive or negative integers. Equation (11.31) is proved by Eqs. (11.35) and (11.37).

$\qquad\qquad\qquad\qquad\qquad\qquad\qquad\qquad\qquad\qquad\qquad\qquad\qquad$ $\square$

11.2.4 *Weierstrass $\wp$ function*

We will now examine Weierstrass $\wp$ (P) function, as an example of the most simple elliptic function.

11.2.4.1 *Weierstrass ζ function*

Weierstrass ζ (zeta) function we will examine first is not a double periodic function and as such, it is not an elliptic function, but it is useful for discussing the properties of elliptic functions.

Let us define

$$\Omega_{m,n} = m\omega_1 + n\omega_2 \qquad (11.38)$$

for any arbitrary integers m, n and two complex numbers ω_1, ω_2 with $\mathrm{Im}(\omega_2/\omega_1) > 0$. We construct a meromorphic function $\zeta(z)$ that has two complex numbers ω_1, ω_2 a first-order pole $\Omega_{m,n}$ and has its principal part $1/(z - \Omega_{m,n})$;

$$\zeta(z) = \frac{1}{z} + \sum_{m,n}{}' \left(\frac{1}{z - \Omega_{m,n}} + \frac{1}{\Omega_{m,n}} + \frac{z}{\Omega_{m,n}{}^2} \right). \qquad (11.39)$$

The sum $\sum_{m,n}{}'$ of the right-hand side is running for all (positive and negative) integer pairs m, n other than $m = n = 0$. The same kind of sum will come up many times afterward and will be defined in this way without further description.

The convergence of Eq. (11.39) can be shown by considering a Taylor expansion around $z = 0$

$$\frac{1}{z - \Omega_{m,n}} = -\frac{1}{\Omega_{m,n}} - \frac{z}{\Omega_{m,n}^2} - \cdots . \tag{11.40}$$

Proof. When $|z| \leq r$, $|\Omega_{m,n}| \geq 2r$ for any arbitrary $r > 0$

$$\left| \frac{1}{z - \Omega_{m,n}} + \frac{1}{\Omega_{m,n}} + \frac{z}{\Omega_{m,n}^2} \right| = \left| \frac{z^2}{\Omega_{m,n}^2 (z - \Omega_{m,n})} \right|$$

$$\leq \frac{r^2}{|\Omega_{m,n}|^2 |\Omega_{m,n}|/2} = \frac{2r^2}{|\Omega_{m,n}|^3}. \tag{11.41}$$

Let the straight line joining two points $n\omega_1$, $n\omega_2$ be g_n (Fig. 11.5). If the length of a vertical line extending down from the origin to g_1 is d, the length of the vertical leg extending down from origin to g_n is nd. Then, for $m, n \geqq 0$,

$$\sum_{0 < m+n \leq N} \frac{1}{|\Omega_{m,n}|^3} \leq \sum_{m=2}^{N} (m-1) \frac{1}{(md)^3} \leq \sum_{m=2}^{N} \frac{1}{d^3 m^2}$$

and the right-hand side converges when $N \to \infty$. The same discussion is valid for other combinations of m, n. Then

$$\sum_{m,n}' \frac{1}{\Omega_{m,n}^3} \tag{11.42}$$

absolutely converges. Thus, Eq. (11.39) uniformly converges in $|z| \leq r$. $\square$

$\zeta(z)$ is not a double periodic function, so it is not an elliptic function, but it has no singular points other than poles in a finite domain on the z plane. $\zeta(z)$ is called Weierstrass ζ (zeta) function.

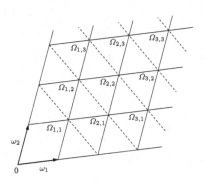

Fig. 11.5 A straight line joining $\Omega_{m,n}$.

11.2.4.2 Definition of $\wp$ function

As already described, there are no elliptic functions with order 1. The elliptic function with order 2 (named Weierstrass $\wp$ function) defined below is one of the simplest elliptic functions.

Definition 11.4. The function defined by the equation

$$\wp(z;\omega_1,\omega_2) = \frac{1}{z^2} + \sum_{m,n}' \left[\frac{1}{(z-\Omega_{m,n})^2} - \frac{1}{{\Omega_{m,n}}^2} \right] \tag{11.43}$$

is called Weierstrass $\wp$ function.

We have already demonstrated that ζ function converges uniformly. Therefore, the termwise differentiation of it is permitted, and this leads to $\wp$ function.

11.2.4.3 Relationship between ζ function and $\wp$ function

$$\frac{\mathrm{d}}{\mathrm{d}z}\zeta(z) = -\wp(z;\omega_1,\omega_2). \tag{11.44}$$

There are many other relationships, but we will not explain them here. Refer to the references [E. T. Whittaker and G. N. Watson (1952)], [A. Hurwitz and R. Courant (1922)].

11.2.4.4 Properties of $\wp$ function

Figure 11.6 shows the behavior of Weierstrass $\wp$ function on the real axis, which has the following properties.

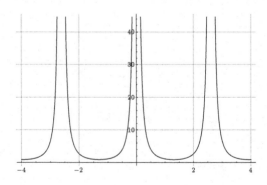

Fig. 11.6 $\wp(z)$ function.

(1) $\wp$ function has the periods ω_1, ω_2. This fact is clear from the definition.

(2) $\wp$ function has the unique second-order pole $z = 0$ within the fundamental period parallelogram, and has no other poles. This fact is also clear from the definition.

From (1) and (2) above, it follows that $\wp$ function is an elliptic function with order 2.

(3) It is clear from the definition that

$$\wp(-z) = \wp(z). \qquad (11.45)$$

(4) For any arbitrary complex number c, there are two points (including multiples) at which $\wp(z) - c = 0$. This is derived from the general properties of elliptic functions.

(5) $\wp(z)$ function satisfies the differential equation

$$(\wp'(z))^2 = 4\wp(z)^3 - g_2\wp(z) + g_3, \qquad (11.46)$$

where

$$g_2 = 60\sideset{}{'}\sum_{m,n} \frac{1}{\Omega_{m,n}{}^4}, \qquad g_3 = 140\sideset{}{'}\sum_{m,n} \frac{1}{\Omega_{m,n}{}^6}.$$

Proof. This proof is a little complicated, so let us do it properly. Let $r = \mathrm{Min}\{|\Omega_{m,n}| \mid m, n \neq 0\}$. When $0 < |z| < r$, $|z/\Omega_{m,n}| < 1$

$$\frac{1}{(z - \Omega_{m,n})^2} - \frac{1}{\Omega_{m,n}{}^2} = \sum_{k=1}^{\infty} \frac{k+1}{\Omega_{m,n}{}^{k+2}} z^k.$$

When k is an odd integer and summing over m, n, they cancel each other out. So the result only remains when k is even and it produces

$$\wp(z) = \frac{1}{z^2} + \sum_{k=1}^{\infty} (2k+1)G_{2k+2}z^{2k} \qquad (11.47\text{a})$$

$$G_k = \sideset{}{'}\sum_{m,n} \frac{1}{\Omega_{m,n}{}^k}.$$

Using G_k defined here,

$$g_2 = 60G_4, \qquad g_3 = 140G_6.$$

Direct differentiation of Eq. (11.47a) produces

$$\wp'(z) = -\frac{2}{z^3} + 6G_4z + 20G_6z^3 + \cdots, \qquad (11.47\text{b})$$

and

$$[\wp'(z)]^2 = \frac{4}{z^6} - 24\frac{G_4}{z^2} - 80G_6 + \cdots \qquad (11.47c)$$

$$[\wp(z)]^3 = \frac{1}{z^6} + 9\frac{G_4}{z^2} + 15G_6 + \cdots \qquad (11.47d)$$

and other relations. Therefore,

$$[\wp'(z)]^2 - 4[\wp(z)]^3 = -60\frac{G_4}{z^2} - 140G_6 + \cdots. \qquad (11.47e)$$

In this way, we obtain

$$[\wp'(z)]^2 - 4[\wp(z)]^3 + g_2\wp(z) = -g_3 + \mathcal{O}(z^2). \qquad (11.47f)$$

Here, $\mathcal{O}(z^2)$ means z^2- and other higher-order terms of z. The right-hand side of Eq. (11.47f) is a holomorphic function around $z = 0$. The left-hand side is an elliptic function with the same fundamental period as $\wp(z)$. According to the basic properties of elliptic functions, the left-hand side of Eq. (11.47f) must be a constant, so the quadratic and higher-order terms of z must identically be 0. ☐

(6) For Weierstrass $\wp(z)$ function, addition theorems are similar to the addition theorem of trigonometric functions, and readers may refer to references [M. Toda (2001)], [A. Hurwitz and R. Courant (1922)].

11.2.5 *Jacobi's elliptic functions* sn w, cn w, dn w

We introduced elliptic integrals of the first kind at the beginning of this chapter. We will now define Jacobi's elliptic function, which may involve some repetition.

11.2.5.1 *Definition of Jacobi's elliptic function*

Definition 11.5. We define the elliptic functions sn (w, k) or sn w as the inverse functions of the following elliptic integral, and k is called the modulus. Here, the rectangular domain on the complex w plane (Fig. 11.3(b)) corresponds to the upper half of the z plane.

$$w = \int_0^z \frac{dz}{\sqrt{(1 - z^2)(1 - k^2z^2)}} = \text{sn}^{-1}z. \qquad (11.48)$$

Sometimes we write

$$w = \text{sn}^{-1}(z, k) \qquad (11.49)$$

to express this explicitly. Taking the inverse of this, we write the elliptic function sn (w, k) or sn w as

$$z = \text{sn}\, w = \text{sn}\,(w, k). \qquad (11.50)$$

Definition 11.6. We define the following functions, which are also Jacobi's elliptic functions:

$$\text{cn}\,(w,k) = \sqrt{1 - \text{sn}^2(w,k)} \tag{11.51a}$$

$$\text{dn}(w,k) = \sqrt{1 - k^2\text{sn}^2(w,k)}. \tag{11.51b}$$

Here, in both these equations, the sign of the root is determined so that this value is positive when w is in $(0, K)$ on the real axis. Figure 11.7 shows the behavior of sn, cn, and dn.

11.2.5.2 *Properties of Jacobi's elliptic functions*

(1) If the sign of the root is determined as above, $\text{cn}\,(w,k)$, $\text{dn}(w,k)$ are single-valued meromorphic functions on the complex w plane.
(2) When $k = 0$,

$$\text{sn}\,(w,0) = \sin w \tag{11.52a}$$

$$\text{cn}\,(w,0) = \cos w \tag{11.52b}$$

$$\text{dn}\,(w,0) = 1. \tag{11.52c}$$

For $k = 1$,

$$\text{sn}\,(w,1) = \tanh w \tag{11.53a}$$

$$\text{cn}\,(w,1) = \text{sech}\ w \tag{11.53b}$$

$$\text{dn}\,(w,1) = \text{sech}\ w. \tag{11.53c}$$

(3) For $z = \text{sn}\,w$, we obtain, from the definition of Eq. (11.48),

$$\frac{dw}{dz} = \frac{1}{\sqrt{(1 - z^2)(1 - k^2z^2)}}. \tag{11.54}$$

It follows that

$$\frac{dz}{dw} = \sqrt{(1 - \text{sn}^2 w)(1 - k^2\text{sn}^2 w)} = \text{cn}\ w\,\text{dn}\ w. \tag{11.55}$$

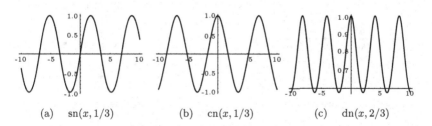

(a) $\text{sn}(x, 1/3)$ (b) $\text{cn}(x, 1/3)$ (c) $\text{dn}(x, 2/3)$

Fig. 11.7   Jacobi's elliptic functions $\text{sn}\,(x, 1/3)$, $\text{cn}\,(x, 1/3)$, $\text{dn}(x, 2/3)$.

Therefore, we obtain

$$\frac{\mathrm{d}\,\mathrm{sn}\,w}{\mathrm{d}w} = \mathrm{cn}\,w\ \mathrm{dn}\,w, \tag{11.56a}$$

and similarly

$$\frac{\mathrm{d}\,\mathrm{cn}\,w}{\mathrm{d}w} = -\mathrm{sn}\,w\,\mathrm{dn}\,w \tag{11.56b}$$

$$\frac{\mathrm{d}\,\mathrm{dn}\,w}{\mathrm{d}w} = -k^2\mathrm{sn}\,w\,\mathrm{cn}\,w. \tag{11.56c}$$

(4) The following property can be demonstrated from the definition.

$$\mathrm{sn}\,(-w) = -\mathrm{sn}\,w. \tag{11.57}$$

(5) The following property was shown at the beginning of this chapter:

$$\mathrm{sn}\,(w + 2K) = -\mathrm{sn}\,w, \tag{11.58a}$$

$$\mathrm{sn}\,(w + 2\mathrm{i}K') = \mathrm{sn}\,w. \tag{11.58b}$$

In a similar way, we can get

$$\mathrm{cn}\,(w + 2K) = -\mathrm{cn}\,w, \qquad \mathrm{cn}\,(w + 2\mathrm{i}K') = -\mathrm{cn}\,w \tag{11.58c}$$

$$\mathrm{dn}(w + 2K) = \mathrm{dn}\,w, \qquad \mathrm{dn}(w + 2\mathrm{i}K') = -\mathrm{dn}\,w. \tag{11.58d}$$

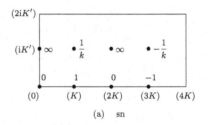

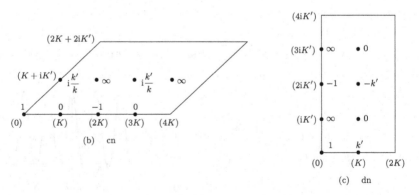

Fig. 11.8 The fundamental period parallelogram of Jacobi's elliptic functions $\mathrm{sn}\,w$, $\mathrm{cn}\,w$, $\mathrm{dn}\,w$, and the values at a number of specific points. $k' = \sqrt{1 - k^2}$.

(6) For the fundamental period of Jacobi's elliptic functions, we demonstrated in Sec. 11.1.5 that the fundamental period of $\operatorname{sn} w$ is $4K$, $2\mathrm{i}K'$.

$$\operatorname{sn}\left(w+4K\right) = \operatorname{sn} w \tag{11.59a}$$
$$\operatorname{sn}\left(w+2\mathrm{i}K'\right) = \operatorname{sn} w. \tag{11.59b}$$

The same is true for cn, dn. Figure 11.8 shows the fundamental period parallelogram for sn, cn, dn, and the values at specific points. These are also summarized in Table 11.1.

Table 11.1 Fundamental periods, zero points, and poles of Jacobi's elliptic functions.

	Fundamental period	Zero point	Pole
$\operatorname{sn} w$	$4K$, $2\mathrm{i}K'$	0, $2K$	$\mathrm{i}K'$, $2K+\mathrm{i}K'$
$\operatorname{cn} w$	$4K$, $2K+2\mathrm{i}K'$	K, $-K$	$\mathrm{i}K'$, $2K+\mathrm{i}K'$
$\operatorname{dn} w$	$2K$, $4\mathrm{i}K'$	$K+\mathrm{i}K'$, $K-\mathrm{i}K'$	$\mathrm{i}K'$, $-\mathrm{i}K'$

There are other addition theorems etc. for Jacobi's elliptic functions, and elliptic functions can reasonably be called the generalization of trigonometric functions. The details can be found in references [M. Toda (2001)], [A. Hurwitz and R. Courant (1922)].

Chapter 12

Ordinary Differential Equations of Complex Variables

In this chapter, we will consider differential equations of a complex variable and their solutions. Specifically, we will examine the functions called hypergeometric functions and confluent hypergeometric functions. These are solutions to differential equations that appear widely in physics and engineering.

12.1 Differential equations and series solutions

In solving differential equations, the scope of what can be achieved with **quadrature** (the method of using a finite number of indefinite integrals to express solutions) is generally very limited. So, widely used solving methods are those, such as the Taylor expansion or the Laurent expansion, which hypothesize solutions in the form of a series.

12.1.1 *nth-order linear ordinary differential equations for complex variables*

Consider linear ordinary differential equations of complex variable z

$$p_0(z)\frac{\mathrm{d}^n}{\mathrm{d}z^n}w(z) + p_1(z)\frac{\mathrm{d}^{n-1}}{\mathrm{d}z^{n-1}}w(z) + \cdots + p_n(z)w(z) = 0. \qquad (12.1)$$

The coefficient $p_k(z)$ is a function of z.

This kind of problem is still meaningful even if a variable is limited to the range of real numbers. In that case, the existence and uniqueness of a solution are guaranteed by Cauchy's theorem under Lipschitz conditions. This is also true locally for the differential equations of a complex variable. In the case of a complex variable when a domain of a variable is extended globally, the question of whether the solution still holds becomes another important problem. In other words, the question is whether it is possible to take the solution obtained around a certain point can be analytically continued over a wider domain on Gauss plane.

12.1.2   *Series solution around a regular point*

Definition 12.1. When all coefficient $p_k(z)$ of the differential equation (12.1) are holomorphic around $z = a$ and $p_0(a) \neq 0$, the point $z = a$ is called a **regular point** or an **ordinary point** of the differential equation (12.1).

Theorem 12.1. *When a point $z = a$ is a regular point of the differential equation (12.1), there is only one holomorphic solution at $z = a$ that satisfies the initial condition*

$$w(a) = b_0, \ w'(a) = b_1, \ \cdots, \ w^{(n-1)}(a) = b_{n-1}. \tag{12.2}$$

Proof. It is cumbersome to present proof in detail for general n, so we will state only the key points.

Assume that a solution takes the form of a Taylor series.

$$w(z) = \sum_{m=0}^{\infty} c_m(z-a)^m. \tag{12.3}$$

Because $w(a) = c_0$, $w'(a) = c_1$, $\cdots$, etc., $c_0, c_1, \cdots, c_{n-1}$ are determined as $c_i = b_i$ $(i = 0, 1, \cdots, n-1)$ from the initial conditions. $p_k(z)$ can also be given by a Taylor expansion around $z = a$ and substituted into Eq. (12.1), obtaining simultaneous equations for c_m $(m = 0, 1, 2, \cdots)$ if the coefficient for each $(z-a)^k$ is set to 0. This set of equations has a structure that c_n is expressed by $c_0, c_1, \cdots, c_{n-1}$, and c_{n+1} by $c_0, c_1, \cdots, c_{n-1}, c_n$ and so on. So once $c_0, c_1, \cdots, c_{n-1}$ are determined from the initial conditions, all other c_m $(m = n, n+1, \cdots)$ are expressed by them, and the solution (12.3) can be obtained. $\square$

Let the solution for the initial condition $w(a) = 1$, $w'(a) = w''(a) = \cdots = 0$ be $w_0(z)$, the solution for the initial condition $w(a) = 0$, $w'(a) = 1$, $w''(a) = \cdots = 0$ be $w_1(z)$, and the solution for the initial condition $w(a) = w'(a) = 0$, $w''(a) = 1$, $w'''(a) = \cdots = 0$ be $w_2(z)$, etc. In this case, we can write the solution which satisfies the initial condition (12.2) as

$$w(a) = b_0 w_0(z) + b_1 w_1(z) + \cdots + b_{n-1} w_{n-1}(z). \tag{12.4}$$

Example 12.1. For Legendre differential equation of order n

$$(1 - z^2)\frac{\mathrm{d}^2 w(z)}{\mathrm{d}z^2} - 2z\frac{\mathrm{d}w(z)}{\mathrm{d}z} + n(n+1)\,w(z) = 0, \tag{12.5}$$

$z = 0$ is a regular point. As performed above, the solution can be found as a Taylor series of z.

12.1.3   *Series solution around a regular singular point*

Now let us examine a case in which one of $p_k(z)$ is not holomorphic at $z = a$. In the following discussion, we take $a = 0$, but it is possible to shift any arbitrary point to the origin 0 with a variable transformation, so generality is not lost.

12.1.3.1   *Regular singular point*

Consider a linear ordinary differential equation for a complex variable z

$$z^n p_0(z)\frac{\mathrm{d}^n}{\mathrm{d}z^n}w(z) + z^{n-1}p_1(z)\frac{\mathrm{d}^{n-1}}{\mathrm{d}z^{n-1}}w(z) + \cdots$$
$$+ zp_{n-1}(z)\frac{\mathrm{d}}{\mathrm{d}z}w(z) + p_n(z)\,w(z) = 0. \qquad (12.6)$$

When $p_k(z)$ ($k = 0, 1, \cdots, n$) are all holomorphic at $z = 0$, and $p_0(0) \neq 0$, $z = 0$ is said to be a **regular singular point** of this differential equation.[1] We know that in this case at least one solution can be written as

$$w(z) = z^\rho \sum_{k=0}^{\infty} c_k z^k \qquad (12.7)$$

with an exponent ρ. Points that are neither regular points nor regular singular points are called **irregular singular points**.

12.1.3.2   *Series solution around a regular singular point: Case of* $n = 2$

In the following, we are discussing a series solution around a regular singular point. To make the discussion clearer, we will take an example of $n = 2$, which is to say, the 2nd order linear ordinary differential equation

$$z^2 p(z)\,w''(z) + zq(z)\,w'(z) + r(z)\,w(z) = 0. \qquad (12.8)$$

---

[1]The origin of the name "regular singular point" is as follows: If a real number μ is chosen appropriately for a regular singular point a of $w(z)$, then $(z - a)^\mu w(z) \to 0$ when $z \to a$. In that case, $w(z)$ either approaches a determinate finite value or approaches ∞, so the limit of $w(z)$ is determined "regularly" in any case.

Here, because of assumption, we can expand them as

$$p(z) = \sum_{k=0}^{\infty} p_k z^k \qquad [p_0 = p(0) \neq 0]$$

$$q(z) = \sum_{k=0}^{\infty} q_k z^k \qquad\qquad (12.9)$$

$$r(z) = \sum_{k=0}^{\infty} r_k z^k.$$

Substituting Eqs. (12.7), (12.9) into Eq. (12.8) produces

$$\sum_{k=0}^{\infty} z^{\rho+k} \big\{ c_k [p_0(\rho+k)^2 + (q_0 - p_0)(\rho+k) + r_0]$$

$$+ c_{k-1}[p_1(\rho+k-1)^2 + (q_1 - p_1)(\rho+k-1) + r_1] + \cdots$$

$$+ c_0\{p_k\rho^2 + (q_k - p_k)\rho + r_k]\}$$

$$= \sum_{k=0}^{\infty} z^{\rho+k}[c_k f_0(\rho+k) + c_{k-1}f_1(\rho+k-1) + \cdots + c_0 f_k(\rho)] = 0 \ .$$

$$(12.10)$$

Here, $f_k(\lambda)$ is a quadratic equation of λ;

$$f_k(\lambda) = p_k \lambda^2 + (q_k - p_k)\lambda + r_k \qquad (k = 0, 1, \cdots). \qquad (12.11)$$

Equation (12.10) is an identity for any value of z, so for $m = 0, 1, 2, \cdots$

$$c_m f_0(\rho + m) + c_{m-1} f_1(\rho + m - 1) + \cdots + c_0 f_m(\rho) = 0 \qquad (12.12)$$

is obtained. For $m = 0$ in particular, taking $c_0 \neq 0$,

$$f_0(\rho) = p_0 \rho^2 + (q_0 - p_0)\rho + r_0 = p_0(\rho - \rho_1)(\rho - \rho_2) = 0 \qquad (12.13)$$

must be satisfied. In other words, we determine the indices $\rho = \rho_1, \rho_2$ by solving Eq. (12.13) with $c_0 = 1$, and determine $c_1, c_2, \cdots$, sequentially by Eq. (12.12). Equation (12.13) is called an **indicial equation**.

12.1.3.3 *The case that, in Eq. (12.13), the difference between the indices ρ_1 and ρ_2 is neither 0 nor integer*

Particular care is required if the indices ρ_1 and ρ_2 in Eq. (12.13) are such that $\rho_1 = \rho_2$ or $\rho_1 - \rho_2 = $ integer. In other cases ($\rho_1 - \rho_2 = $ non-integer number), taking $c_0 = 1$, two independent solutions $w_i(z)$ ($i = 1, 2$), taking the form of Eq. (12.7), are determined for the two exponents ρ_i ($i = 1, 2$).

$$w_i(z) = z^{\rho_i} \sum_{k=0}^{\infty} c_k(\rho_i) z^k \qquad (i = 1, 2). \qquad (12.14)$$

The general solution is the linear combination of these two.

Example 12.2 (Bessel differential equation of order ν).

Consider the series solution of Bessel differential equation of order ν

$$w''(z) + \frac{1}{z}w'(z) + \left(1 - \frac{\nu^2}{z^2}\right)w(z) = 0 \qquad (12.15)$$

around $z = 0$. In that case, $z = 0$ is a regular singular point, so we can hypothesize a solution of

$$w(z) = z^\rho \sum_{m=0}^{\infty} c_m z^m. \qquad (12.16)$$

The indicial equation is $\rho^2 - \nu^2 = 0$, so

$$\rho = \pm\nu \qquad (12.17)$$

are the exponents to find. The recurrence formula which determines each coefficient c_m is

$$c_m = \frac{c_{m-2}}{\nu^2 - (\rho + m)^2} = -\frac{c_{m-2}}{(\rho + m - \nu)(\rho + m + \nu)}. \qquad (12.18)$$

Provided $\rho_1 - \rho_2 = 2\nu$ is not an integer, the coefficients are determined by Eq. (12.18) for either value of $\rho = \pm\nu$.

$$c_m = -\frac{c_{m-2}}{m(\pm 2\nu + m)} = \frac{c_{m-4}}{m(m-2)(\pm 2\nu + m)(\pm 2\nu + m - 2)} = \cdots \qquad (12.19)$$

so c_m can be expressed by c_0 or c_1. The equation which determines c_1 is Eq. (12.12), with $m = 1$, so $c_1 f(\pm\nu + 1) + c_0 f(\pm\nu) = c_1 f(\pm\nu + 1) = 0$. Thus,

$$c_1 = 0. \qquad (12.20)$$

Therefore,

$$c_{2m-1} = 0 \qquad (12.21a)$$

$$c_{2m} = c_0 \frac{(-1)^m}{2^{2m} \, m!(\pm\nu + 1)(\pm\nu + 2)\cdots(\pm\nu + m)}. \qquad (12.21b)$$

When $\nu = n$ (integer), these two solutions are linearly dependent, only differing by a constant factor, so we must use another method. That is Frobenius' method which is the main problem in the next section.[2]

---

[2]In the case of Bessel's function, the coefficients of odd-numbered orders, and those of even-numbered orders, are determined independently, so in practice, we can proceed in this way even if ν is a half-odd integer. Frobenius' method must be followed if $\nu = n$ (0 or an integer).

12.1.4 Frobenius' method for solution around regular singular point

If the indicial equation has multiple roots, or if the difference between the two roots of the indicial equation is an integer, the second solution cannot be obtained by the method described in the preceding section. Frobenius' method can give the second solution in this case. The second solution is generally of the form $\log z \times$ (an infinite series of z) + (an infinite series z).

Let us consider the solution

$$w(z) = z^{\rho_i} \sum_{k=0}^{\infty} c_k(\rho_i) z^k \tag{12.22}$$

in the same way as Eqs. (12.7) or (12.14). In that case, we obtain

$$c_m f_0(\rho + m) + c_{m-1} f_1(\rho + m - 1) + \cdots + c_0 f_m(\rho) = 0 \tag{12.23}$$

in the same way as Eq. (12.12).

Then factorizing the indicial equation that we obtained to find the two roots ρ_1, ρ_2 ($\rho_1 \geq \rho_2$), we can write

$$f_0(\rho) = p_0(\rho - \rho_1)(\rho - \rho_2). \tag{12.24}$$

12.1.4.1 The case $f_0(\rho) = p_0(\rho - \rho_0)^2$, i.e. $\rho_1 = \rho_2 = \rho_0$

In this case, the procedure explained previously can be followed where one of the solutions is $\rho = \rho_0$. Choosing $c_0(\rho_0) = 1$, use Eq. (12.23) and

$$c_m(\rho_0) f_0(\rho_0 + m) + c_{m-1}(\rho_0) f_1(\rho_0 + m - 1) + \cdots + c_0(\rho_0) f_m(\rho_0) = 0$$
$$(m = 1, 2, \cdots) \tag{12.25}$$

to determine $c_m(\rho_0)$ ($m = 1, 2, \cdots$). The series (12.22), taking this $c_m(\rho_0)$ as a coefficient, becomes

$$w(z, \rho_0) = w_1(z) \equiv z^\rho \sum_{m=0}^{\infty} c_m(\rho_0) z^m \tag{12.26}$$

which is the first solution.

How should we go about finding the second solution? In this case, keeping ρ as a variable, we can determine $c_m(\rho)$ ($m = 1, 2, \cdots, m \neq 0$), [as functions of ρ and $c_0(\rho)$], by Eq. (12.23)

$$c_m(\rho) f_0(\rho + m) + c_{m-1}(\rho) f_1(\rho + m - 1) + \cdots$$
$$+ c_0(\rho) f_m(\rho) = 0 \qquad (m = 1, 2, \cdots).$$

Substituting $c_m(\rho)$ $(m = 1, 2, \cdots)$ determined in this way back into the original differential equation (12.8) produces

$$z^2 p(z) w''(z) + z q(z) w'(z) + r(z) w(z) = c_0(\rho) f_0(\rho) z^\rho. \qquad (12.27)$$

Using this series, it *a priori*, the second solution may be taken as

$$w_2(z) = \lim_{\rho \to \rho_0} \frac{\partial}{\partial \rho} w(z, \rho). \qquad (12.28)$$

Proof of Eq. (12.28). Since the differential operators $\partial/\partial\rho$ and $\mathrm{d}/\mathrm{d}z$ are commutative, it is clear that the equation

$$\frac{\partial}{\partial \rho} \left[z^2 p(z) \frac{\mathrm{d}^2}{\mathrm{d}z^2} + z q(z) \frac{\mathrm{d}}{\mathrm{d}z} + r(z) \right] w(z, \rho)$$
$$= \left[z^2 p(z) \frac{\mathrm{d}^2}{\mathrm{d}z^2} + z q(z) \frac{\mathrm{d}}{\mathrm{d}z} + r(z) \right] \frac{\partial}{\partial \rho} w(z, \rho) \qquad (12.29)$$

is true for Eq. (12.8). Furthermore, if we use Eq. (12.27), and choose $c_0(\rho_0) = c_0 = $ (constant), we can show the following equation

$$\lim_{\rho \to \rho_0} \frac{\partial}{\partial \rho} \left[z^2 p(z) \frac{\mathrm{d}^2}{\mathrm{d}z^2} + z q(z) \frac{\mathrm{d}}{\mathrm{d}z} + r(z) \right] w(z, \rho) = \lim_{\rho \to \rho_0} \frac{\partial}{\partial \rho} c_0 f_0(\rho) z^\rho$$
$$= c_0 p_0 \lim_{\rho \to \rho_0} \frac{\partial}{\partial \rho} (\rho - \rho_0)^2 z^\rho$$
$$= 0. \qquad (12.30)$$

According to Eqs. (12.29) and (12.30), we can see that the function $w_2(z)$ in Eq. (12.28) is the solution for the original differential equation. $\qquad \square$

Let us calculate using Eq. (12.28). Using a relation

$$\frac{\partial}{\partial \rho} z^\rho = \frac{\partial}{\partial \rho} e^{\rho \log z} = \log z \cdot e^{\rho \log z} = \log z \cdot z^\rho$$

we find, from Eq. (12.26),

$$w_2(z) = \log z \cdot \sum_{m=0}^{\infty} c_m(\rho_0) z^{\rho_0 + m} + z^{\rho_0} \sum_{m=0}^{\infty} \left[\frac{\mathrm{d}c_m(\rho)}{\mathrm{d}\rho} \right]_{\rho = \rho_0} z^m, \qquad (12.31)$$

and because $(\mathrm{d}/\mathrm{d}\rho) c_0(\rho) = 0$, due to our choice of $c_0(\rho) = 1$ (constant), we obtain

$$w_2(z) = \log z \cdot w_1(z) + z^{\rho_0} \sum_{m=1}^{\infty} \left[\frac{\mathrm{d}c_m(\rho)}{\mathrm{d}\rho} \right]_{\rho = \rho_0} z^m. \qquad (12.32)$$

Here, $c_m(\rho)$ $(m = 1, 2, \cdots)$ is determined by Eq. (12.25). This is the second solution we are looking for.

12.1.4.2 *The case $f_0(\rho) = p_0(\rho - \rho_1)(\rho - \rho_1 + n)$, i.e. $\rho_1 - \rho_2 = n > 0$ (n: an integer)*

For the first index $\rho = \rho_1$, we take $c_0 = 1$, in exactly the same way as before, and c_m ($m = 1, 2, \cdots$) are determined uniquely by Eq. (12.12). From this, the first solution is determined as

$$w_1(z) = z^{\rho_1} \sum_{m=0}^{\infty} c_m(\rho_1) z^m. \tag{12.33}$$

For the second index $\rho = \rho_2$, and the case of $m = n$ in Eq. (12.23), one should separately consider the case where the following equation holds and the case it does not hold:

$$c_{n-1} f_1(\rho_2 + n - 1) + c_{n-2} f_2(\rho_2 + n - 2) + \cdots + c_0 f_n(\rho_2) = 0. \tag{12.34}$$

(1) **The case of $c_{n-1} f_1(\rho_2 + n - 1) + c_{n-2} f_2(\rho_2 + n - 2) + \cdots + c_0 f_n(\rho_2) \neq 0$.**
In this case, $c_n(\rho_2)$ cannot be obtained from Eq. (12.12)

$$c_m f_0(\rho + m) + c_{m-1} f_1(\rho + m - 1) + \cdots + c_0 f_m(\rho) = 0.$$

The reason is as follows:
If we set $\rho = \rho_2$ and $m = n$ in Eq. (12.12), we get

$$c_n f_0(\rho_2 + n) + \{ c_{n-1} f_1(\rho_2 + n - 1) + c_{n-2} f_1(\rho_2 + n - 2) + \cdots$$
$$+ c_0 f_n(\rho_2) \} = 0, \tag{12.35}$$

but Eq. (12.34) does not hold (from the assumption), and

$$f_0(\rho_2 + n) = f_0(\rho_1) = 0.$$

Therefore, Eq. (12.35) is not true, and c_n can not be determined.
However, since c_0 can be chosen arbitrarily, let us assume with a constant c_0'

$$c_0(\rho) = (\rho - \rho_2) c_0' = (\rho - \rho_1 + n) c_0', \qquad c_0' = \left[\frac{d}{d\rho} c_0(\rho) \right]_{\rho = \rho_2}. \tag{12.36}$$

After that, we again write the coefficient as c_m determined by Eq. (12.12). Writing more explicitly, we get the set of equations

$$\{ (\rho - \rho_2) c_0' \} f_0(\rho) = 0$$

$$c_1 f_0(\rho + 1) + \{ (\rho - \rho_2) c_0' \} f_1(\rho) = 0$$

$$\vdots$$

$$c_{n-1} f_0(\rho + n - 1) + c_{n-2} f_1(\rho + n - 2) + \cdots + \{ (\rho - \rho_2) c_0' \} f_{n-1}(\rho) = 0$$

$$c_n f_0(\rho + n) + c_{n-1} f_1(\rho + n - 1) + \cdots + \{ (\rho - \rho_2) c_0' \} f_n(\rho) = 0$$

$$\vdots$$

$$c_m f_0(\rho + m) + c_{m-1} f_1(\rho + m - 1) + \cdots + \{ (\rho - \rho_2) c_0' \} f_m(\rho) = 0$$

$$\vdots$$

then take the limit of $\rho \to \rho_2$ and determine c_m. After all, we get, making c_0' an arbitrary constant,

$$c_0 = \lim_{\rho \to \rho_2} (\rho - \rho_2)c_0' = 0 \qquad (12.37\text{a})$$

$$c_1 = c_2 = \cdots = c_{n-1} = 0 \qquad (12.37\text{b})$$

$$(c_n \text{ is indefinite})$$

$$c_{n+1}f_0(\rho + n + 1) + c_n f_1(\rho + n) = 0 \qquad (12.37\text{c})$$

$$\vdots$$

$$c_m f_0(\rho + m) + c_{m-1}f_1(\rho + m - 1) + \cdots$$
$$+ c_n f_{m-n}(\rho + n) = 0 \qquad (12.37\text{d})$$

$$\vdots \quad .$$

If we choose $c_n = 1$, the set of equations (12.37c)–(12.37d) exactly coincide with that for determining the coefficients of the first solution $w_1(z)$. Choosing $c_n = 1$ and determining the coefficients $c_m(\rho)$ ($m \geq n + 1$) by Eqs. (12.37c)–(12.37d), we obtain the series as

$$w(z, \rho) = z^\rho \sum_{m=0}^{\infty} c_{n+m}(\rho)z^m. \qquad (12.38)$$

The second solution is

$$w_2(z) = \lim_{\rho \to \rho_2} \frac{\partial}{\partial \rho} w(z, \rho), \qquad (12.39)$$

as can be shown in the same way as in the proof of Eq. (12.28). This second solution can be written more explicitly as

$$w_2(z) = \log z \cdot w(z, \rho_1) + z^{\rho_2} \sum_{m=0}^{\infty} \left[\frac{dc_{n+m}(\rho)}{d\rho} \right]_{\rho=\rho_2} z^m, \qquad (12.40)$$

where $w(z, \rho_1)$ is just Eq. (12.38), and is only different from the first solution (12.33) by at most a constant factor. The coefficients c_{n+m} used here, and the coefficients c_m of the first solution in $w_1(z)$ in Eq. (12.33) only differ by this constant factor.

(2) **The case of** $c_{n-1}f_1(\rho_2+n-1)+c_{n-2}f_2(\rho_2+n-2)+\cdots+c_0 f_n(\rho_2) = 0$. This case is probably clear from the explanations so far. If Eq. (12.34) is true, once we determine $c_1, \cdots, c_{n-1}$ with the coefficient $c_0 = 1$, then we can let $c_n = 0$. Furthermore, we can determine the further c_m ($m = n + 1, n + 2, \cdots$) by Eq. (12.12). Therefore, we have determined all c_m by this method, and get a second solution

$$w_2(z) = z^{\rho_2} \sum_{m=0}^{\infty} c_m(\rho_2)z^m. \qquad (12.41)$$

Thus, we obtain a second solution that does not have the $\log z$ term.

Example 12.3 (Bessel function of integer order). In Example 12.2, we discussed Bessel differential equation of non-integer order ν. We will now consider the series solution around the regular singular point $z = 0$ of Bessel differential equation of an integer-order

$$w''(z) + \frac{1}{z}w'(z) + \left(1 - \frac{n^2}{z^2}\right)w(z) = 0. \tag{12.42}$$

In this case, the index determined by the indicial equation is $\rho = \pm n$. Following Frobenius' method, the two solutions are determined as

$$J_n(z) = \left(\frac{z}{2}\right)^n \sum_{m=0}^{\infty} \frac{(-1)^m (z/2)^{2m}}{m!(n+m)!} \tag{12.43a}$$

$$w_2(z) = \{2\log z + \phi(n-1)\}J_n(z) - \hat{w}_2(z), \tag{12.43b}$$

where,

$$\hat{w}_2(z) = \left(\frac{z}{2}\right)^{-n} \sum_{m=0}^{n-1} \frac{(n-m-1)!}{m!}\left(\frac{z}{2}\right)^{2m}$$

$$+ \left(\frac{z}{2}\right)^n \sum_{m=0}^{\infty} \frac{(-1)^m}{m!(n+m)!}\{\phi(m) + \phi(n+m)\}\left(\frac{z}{2}\right)^{2m}$$

$$\phi(m) = \left(\frac{1}{1} + \frac{1}{2} + \frac{1}{3} + \cdots + \frac{1}{m}\right).$$

Here, $J_n(z)$ of Eq. (12.43a) is Bessel function of order n. For the second solution, we usually use a linear combination of $J_n(z)$ and $w_2(z)$, rather than $w_2(z)$ in Eq. (12.43b).

Their details will be described in Sec. 15.3.

12.1.5 *Series solution around the point at infinity*

Consider a series solution around $z = \infty$. Perform the variable transformation $\zeta = 1/z$ on the original differential equation and consider the series solution around $\zeta = 0$.

When $\zeta = 0$ is a regular point, we can obtain two independent solutions in the form of the Taylor series

$$w(\zeta) = \sum_{m=0}^{\infty} c_m \zeta^m. \tag{12.44a}$$

Substituting back into z produces

$$w(z) = \sum_{m=0}^{\infty} c_m z^{-m}. \tag{12.44b}$$

When $\zeta = 0$ is a regular singular point, we can assume a solution

$$w(\zeta) = \sum_{m=0}^{\infty} c_m \zeta^{\rho+m} \tag{12.44c}$$

as a solution. We can substitute this into the differential equation rewritten for ζ and repeat the same discussion as that employed up to now.

Let us consider the differential equation

$$w''(z) + P(z)\, w'(z) + Q(z)\, w(z) = 0 \tag{12.45}$$

more concretely. Performing $\zeta = 1/z$ produces

$$\frac{d}{dz} = -\zeta^2 \frac{d}{d\zeta}, \quad \frac{d^2}{dz^2} = -\zeta^2 \frac{d}{d\zeta} \cdot \left(-\zeta^2 \frac{d}{d\zeta}\right) = \zeta^4 \frac{d^2}{d\zeta^2} + 2\zeta^3 \frac{d}{d\zeta},$$

so Eq. (12.45) becomes

$$\frac{d^2}{d\zeta^2} w(\zeta) + \left[\frac{2}{\zeta} - \frac{1}{\zeta^2} P\left(\frac{1}{\zeta}\right)\right] \frac{d}{d\zeta} w(\zeta) + \frac{1}{\zeta^4} Q\left(\frac{1}{\zeta}\right) w(\zeta) = 0. \tag{12.46}$$

Once we understand the behavior of the coefficients of Eq. (12.46)

$$\left[\frac{2}{\zeta} - \frac{1}{\zeta^2} P\left(\frac{1}{\zeta}\right)\right], \quad \frac{1}{\zeta^4} Q\left(\frac{1}{\zeta}\right)$$

around $\zeta = 0$, we can see whether $\zeta = 0$ ($z = \infty$) is a regular point, a regular singular point, or an irregular singular point. We can also obtain the solution or investigate its properties.

Assume $P(1/\zeta)$, $Q(1/\zeta)$ can be expanded as

$$P\left(\frac{1}{\zeta}\right) = P_0 + P_1\zeta + P_2\zeta^2 + \cdots \tag{12.47a}$$

$$Q\left(\frac{1}{\zeta}\right) = Q_0 + Q_1\zeta + Q_2\zeta^2 + Q_3\zeta^3 + \cdots. \tag{12.47b}$$

12.1.5.1 *Conditions for $z = \infty$ to be a regular singular point*

The condition for $z = \infty$ ($\zeta = 0$) to be a regular singular point is, from the definition, that two functions $P(1/\zeta)$ and $Q(1/\zeta)$ can be expanded as follows:

$$P\left(\frac{1}{\zeta}\right) = P_1\zeta + P_2\zeta^2 + \cdots \tag{12.48a}$$

$$Q\left(\frac{1}{\zeta}\right) = Q_2\zeta^2 + Q_3\zeta^3 + \cdots. \tag{12.48b}$$

12.1.5.2 *Conditions for $z = \infty$ to be a regular point*

The condition for $z = \infty$ ($\zeta = 0$) to be a regular point is that the two functions can be expressed as

$$P\left(\frac{1}{\zeta}\right) = 2\zeta + P_2\zeta^2 + \cdots \tag{12.49a}$$

$$Q\left(\frac{1}{\zeta}\right) = Q_4\zeta^4 + Q_5\zeta^5 + \cdots. \tag{12.49b}$$

12.2 Riemann's P function and Gauss hypergeometric functions

Our discussion so far has been based on the idea of finding the solution of an ordinary differential equation with the complex variable z in the form of a series. This problem can be discussed from a slightly broader and more general perspective.

12.2.1 *Fuchsian differential equations and Riemann's P function*

12.2.1.1 *Fuchsian differential equations and Fuchsian relation*

A differential equation with only a finite number of regular singular points as singularities is called Fuchsian differential equation.

Consider a 2nd-order linear ordinary differential equation

$$L[w] \equiv \frac{d^2w(z)}{dz^2} + P(z)\frac{dw(z)}{dz} + Q(z)\,w(z) = 0 \tag{12.50}$$

and let $a_1, a_2, \cdots, a_m$ be regular singular points in a finite domain. In that case, $P_1(z)$ is at most an $(m-1)$-th order polynomial of z, while $Q_1(z)$ is at most a $(2m-2)$-th order polynomial of z, so $P(z)$, $Q(z)$ can be written as

$$P(z) = \frac{P_1(z)}{(z - a_1)\cdots(z - a_m)} \tag{12.51a}$$

$$Q(z) = \frac{Q_1(z)}{(z - a_1)^2\cdots(z - a_m)^2}. \tag{12.51b}$$

Then, we can resolve the rational functions (12.51a), (12.51b) into partial fractions, and write them as

$$P(z) = \sum_{k=1}^{m} \frac{A_k}{z - a_k} \tag{12.52a}$$

$$Q(z) = \sum_{k=1}^{m} \left\{ \frac{B_k}{(z - a_k)^2} + \frac{C_k}{z - a_k} \right\}. \tag{12.52b}$$

In this case,

$$\sum_{k=1}^{m} C_k = 0 \tag{12.52c}$$

must be satisfied. Otherwise, $Q_1(z)$ becomes a $(2m-1)$-th order polynomial of z, and this contradicts the fact that $z = \infty$ is a regular point (or a regular singular point).

Also, if we write the index here as ρ_k, the indicial equation for $z = a_k$ $(k = 1, 2, \cdots, m)$ becomes

$$\rho_k(\rho_k - 1) + A_k \rho_k + B_k = 0 \qquad (k = 1, 2, \cdots, m). \tag{12.53}$$

If we let the two roots of Eq. (12.53) be $\rho_k^{(1)}$, $\rho_k^{(2)}$, then we have

$$\rho_k^{(1)} + \rho_k^{(2)} = 1 - A_k, \qquad \rho_k^{(1)} \cdot \rho_k^{(2)} = B_k \qquad (k = 1, 2, \cdots, m). \tag{12.54}$$

If $z = \infty$ is a regular singular point, we can expand $P(z)$, $Q(z)$ around $z = \infty$, to write

$$P(z) = \frac{\sum_{k=1}^{m} A_k}{z} + \frac{\sum_{k=1}^{m} a_k A_k}{z^2} + \cdots \tag{12.55a}$$

$$Q(z) = \frac{\sum_{k=1}^{m} C_k}{z} + \frac{\sum_{k=1}^{m} (a_k C_k + B_k)}{z^2} + \frac{\sum_{k=1}^{m} (a_k{}^2 C_k + 2a_k B_k)}{z^3} + \cdots. \tag{12.55b}$$

Therefore the indicial equation at $z = \infty$ is

$$\rho_\infty(\rho_\infty + 1) - \left(\sum_{k=1}^{m} A_k \right) \rho_\infty + \sum_{k=1}^{m} (a_k C_k + B_k) = 0. \tag{12.56a}$$

In this case,

$$\rho_\infty^{(1)} + \rho_\infty^{(2)} = \left(\sum_{k=1}^{m} A_k \right) - 1, \qquad \rho_\infty^{(1)} \cdot \rho_\infty^{(2)} = \sum_{k=1}^{m} (a_k C_k + B_k) \tag{12.56b}$$

in the same way as Eq. (12.54). Furthermore, from Eqs. (12.54) and (12.56b), we can show

$$\sum_{k=1}^{m} (\rho_k^{(1)} + \rho_k^{(2)}) + \rho_\infty^{(1)} + \rho_\infty^{(2)} = m - 1. \tag{12.56c}$$

Equation (12.56c) is called **Fuchsian relation**. Note that this relation holds in both cases that $z = \infty$ is a regular singular point and a regular point.

12.2.1.2 *When there are only three regular singular points in a finite domain: Papperitz's relation and Riemann's P Function*

Comparing Eqs. (12.55a) and (12.55b) of $P(z)$ and $Q(z)$ with Eqs. (12.49a) and (12.49b), and adding Eq. (12.52c), we obtain the condition for $z = \infty$ to be a regular point as

$$\sum_{k=1}^{m} A_k = 2 \tag{12.57a}$$

$$\sum_{k=1}^{m} (a_k C_k + B_k) = 0 \tag{12.57b}$$

$$\sum_{k=1}^{m} (a_k{}^2 C_k + 2a_k B_k) = 0. \tag{12.57c}$$

In particular, let us consider the case in which a_1, a_2, a_3 are the only three regular singular points and $z = \infty$ is a regular point.[3] The differential equation is

$$\frac{d^2 w(z)}{dz^2} + \left(\sum_{k=1}^{3} \frac{A_k}{z - a_k} \right) \frac{dw(z)}{dz} + \sum_{k=1}^{3} \left[\frac{B_k}{(z - a_k)^2} + \frac{C_k}{z - a_k} \right] w(z) = 0. \tag{12.58}$$

From Eqs. (12.52c), (12.57b) and (12.57c), if we express $\{C_k\}$ using only $\{a_k\}$ and $\{B_k\}$, and use them to rewrite the coefficient of $w(z)$ in Eq. (12.58),[4] we obtain

$$\frac{d^2 w(z)}{dz^2} + \left(\frac{A_1}{z - a_1} + \frac{A_2}{z - a_2} + \frac{A_3}{z - a_3} \right) \frac{dw(z)}{dz}$$
$$+ \left[\frac{B_1(a_1 - a_2)(a_1 - a_3)}{(z - a_1)^2(z - a_2)(z - a_3)} + \frac{B_2(a_2 - a_3)(a_2 - a_1)}{(z - a_2)^2(z - a_3)(z - a_1)} \right.$$
$$\left. + \frac{B_3(a_3 - a_1)(a_3 - a_2)}{(z - a_3)^2(z - a_1)(z - a_2)} \right] w(z) = 0.$$

---

[3]Later, we will discuss the case in which one of the three regular singular points is the point at infinity, as the limit of $a_3 \to \infty$.

[4]It is not difficult to rewrite this, but the manual calculation requires performing a long and awkward calculation. It would be good to use computer software etc. for this.

If we rewrite the differential equation using Eq. (12.54) and letting $m = 3$, it becomes

$$\frac{d^2w(z)}{dz^2} + \left[\sum_{k=1}^{3} \frac{1 - \rho_k^{(1)} - \rho_k^{(2)}}{z - a_k}\right] \frac{dw(z)}{dz} + \left[\frac{\rho_1^{(1)} \rho_1^{(2)}(a_1 - a_2)(a_1 - a_3)}{(z - a_1)^2(z - a_2)(z - a_3)}\right.$$
$$+ \frac{\rho_2^{(1)} \rho_2^{(2)}(a_2 - a_3)(a_2 - a_1)}{(z - a_2)^2(z - a_3)(z - a_1)} + \left.\frac{\rho_3^{(1)} \rho_3^{(2)}(a_3 - a_1)(a_3 - a_2)}{(z - a_3)^2(z - a_1)(z - a_2)}\right] w(z) = 0.$$
$$(12.59)$$

Also, if we take the summation of the first equation of (12.54), and use condition (12.57a), we obtain

$$\rho_1^{(1)} + \rho_1^{(2)} + \rho_2^{(1)} + \rho_2^{(2)} + \rho_3^{(1)} + \rho_3^{(2)} = 1 \qquad (12.60)$$

for the Fuchsian relation (12.56c). Equation (12.59) is called Papperitz's differential equation.

The important fact that we learn from Eq. (12.59) is that if we set the positions a_1, a_2, a_3 of the regular singular points and the series solution indices $\rho_1^{(1)}$, $\rho_1^{(2)}$, $\rho_2^{(1)}$, $\rho_2^{(2)}$, $\rho_3^{(1)}$, $\rho_3^{(2)}$ around those points so that the Fuchsian relation (12.60) is satisfied, Papperitz's differential equation (12.59) is uniquely determined. It is convenient to write the set of general solutions of this differential equation as

$$w(z) = P \left\{ \begin{array}{ccc} a_1 & a_2 & a_3 \\ \rho_1^{(1)} & \rho_2^{(1)} & \rho_3^{(1)} & z \\ \rho_1^{(2)} & \rho_2^{(2)} & \rho_3^{(2)} \end{array} \right\}. \qquad (12.61)$$

This presentation is called **Riemann scheme**, and this function is called Riemann P function.

12.2.1.3 When a_1, a_2, ∞ are regular singular points

If we take $a_3 \to \infty$ in Papperitz's equation (12.59), we obtain

$$\frac{d^2w(z)}{dz^2} + \left[\frac{1 - \rho_1^{(1)} - \rho_1^{(2)}}{z - a_1} + \frac{1 - \rho_2^{(1)} - \rho_2^{(2)}}{z - a_2}\right] \frac{dw(z)}{dz} + \left[\frac{\rho_1^{(1)} \rho_1^{(2)}(a_1 - a_2)}{(z - a_1)^2(z - a_2)}\right.$$
$$+ \frac{\rho_2^{(1)} \rho_2^{(2)}(a_2 - a_1)}{(z - a_2)^2(z - a_1)} + \left.\frac{\rho_\infty^{(1)} \rho_\infty^{(2)}}{(z - a_1)(z - a_2)}\right] w(z) = 0. \qquad (12.62)$$

Readers may feel this discussion somewhat strange, because Papperitz's equation (12.59) is derived from the hypothesis that $z = \infty$ is a regular point, and that the regular singular points a_1, a_2, a_3 locate in a finite domain. In fact, we should let $m = 2$ in Eqs. (12.51a) and (12.51b), and make obvious use of the fact that a_1, a_2, ∞ are regular singular points for this derivation. The result is the same as if we had performed the above kind of limit operation (readers should check this for themselves). This (set of) general solutions for differential equations is

$$w(z) = P \left\{ \begin{array}{ccc} a_1 & a_2 & \infty \\ \rho_1^{(1)} & \rho_2^{(1)} & \rho_\infty^{(1)} & z \\ \rho_1^{(2)} & \rho_2^{(2)} & \rho_\infty^{(2)} \end{array} \right\} \tag{12.63}$$

if we employ Riemann scheme.

12.2.2 Gauss hypergeometric differential equations and hypergeometric functions

12.2.2.1 Gauss hypergeometric differential equations

In Eq. (12.62), the case when $a_1 = 0$, $a_2 = 1$, $\rho_1^{(1)} = 0$, $\rho_1^{(2)} = 1 - \gamma$, $\rho_2^{(1)} = 0$, $\rho_2^{(2)} = \gamma - \alpha - \beta$, $\rho_\infty^{(1)} = \alpha$, and $\rho_\infty^{(2)} = \beta$ is highly important. Substituting them into Eqs. (12.62) and (12.63), we obtain

$$z(1-z)\frac{\mathrm{d}^2 w(z)}{\mathrm{d}z^2} + \{\gamma - (\alpha + \beta + 1)z\}\frac{\mathrm{d}w(z)}{\mathrm{d}z} - \alpha\beta w(z) = 0. \tag{12.64}$$

This equation is particularly called Gauss hypergeometric differential equation.

Using Riemann scheme of Eq. (12.63), the general solution of hypergeometric differential equations can be written as

$$w(z) = P \left\{ \begin{array}{ccc} 0 & 1 & \infty \\ 0 & 0 & \alpha & z \\ 1 - \gamma & \gamma - \alpha - \beta & \beta \end{array} \right\}. \tag{12.65}$$

This differential equation leads to various differential equations in physics and engineering, depending on how α and β are selected. Specific applied examples are presented in Chapters 14 and 15. Let us find solutions to Gauss hypergeometric differential equation specifically expressed in the series expansion.

12.2.2.2 Gauss hypergeometric function I. The solution around the regular singular point $z = 0$

According to the general method for a series solution, we assume

$$w(z) = z^\rho \sum_{m=0}^{\infty} c_m z^m \tag{12.66}$$

as the solution around $z = 0$. The indicial equation is

$$\rho(\rho - 1 + \gamma) = 0, \tag{12.67}$$

so the exponents are

$$\rho = 0, \qquad \rho = 1 - \gamma. \tag{12.68}$$

For the coefficient c_n, we obtain the recurrence formula

$$c_n = \frac{(\rho + n - 1 + \alpha)(\rho + n - 1 + \beta)}{(\rho + n)(\rho + n - 1 + \gamma)} c_{n-1} = \cdots = \frac{[\rho + \alpha]_n [\rho + \beta]_n}{[\rho + 1]_n [\rho + \gamma]_n} c_0. \tag{12.69}$$

Here, we used Pochhammer symbol

$$[a]_0 = 1, \quad [a]_n = a(a + 1)(a + 2) \cdots (a + n - 1). \tag{12.70}$$

If γ is not an integer, we can obtain from the above one of the series solutions around the regular singular point $z = 0$, corresponding to $\rho = 0$, as

$$w_1(z) = \sum_{n=0}^{\infty} \frac{[\alpha]_n [\beta]_n}{n! [\gamma]_n} z^n = 1 + \frac{\alpha \beta}{1 \cdot \gamma} z + \frac{\alpha(\alpha + 1)\beta(\beta + 1)}{1 \cdot 2 \cdot \gamma(\gamma + 1)} z^2 + \cdots . \tag{12.71}$$

This power series solution is called Gauss hypergeometric series, and is written as

$$w_1(z) = F(\alpha, \beta; \gamma; z) = F(\beta, \alpha; \gamma; z). \tag{12.72}$$

Alternatively, Gauss hypergeometric series $F(\alpha, \beta; \gamma; z)$ is written as $F(\alpha, \beta; \gamma | z)$ or ${}_2F_1(\alpha, \beta; \gamma; z)$.

Taking the ratio of the coefficients of the series for the first solution found here, we obtain

$$\lim_{n \to \infty} \left| \frac{c_n}{c_{n+1}} \right| = \lim_{n \to \infty} \left| \frac{(\gamma + n)(n + 1)}{(\alpha + n)(\beta + n)} \right| = 1. \tag{12.73}$$

Therefore, the convergence radius of the hypergeometric series $F(\alpha, \beta; \gamma; z)$ is 1, and $F(\alpha, \beta; \gamma; z)$ is holomorphic when $|z| < 1$. This is called Gauss hypergeometric function, or simply hypergeometric function.

Let us find the second solution $w_2(z)$ in the case where γ is not an integer. Substituting $\rho = 1 - \gamma$ into Eq. (12.69), we get

$$c_n = \frac{[\alpha - \gamma + 1]_n [\beta - \gamma + 1]_n}{n! [2 - \gamma]_n} c_0. \tag{12.74}$$

From that, we can see, by employing the notation $F(\alpha, \beta; \gamma; z)$ again, that the second solution is

$$w_2(z) = z^{1-\gamma} F(\alpha - \gamma + 1, \beta - \gamma + 1; 2 - \gamma; z). \tag{12.75}$$

When γ is an integer, Frobenius' method can be used to find two independent solutions. We will present just the result for that method (for details, refer to pp. 78–86 of Ref. [T. Inui (1962)]). Taking m as 0 or as a positive integer, and defining

$$F_1(\alpha, \beta; m+1; z) = (-1)^{m+1} m! z^{-m} \sum_{k=0}^{m-1} \frac{(-1)^k (m-k-1)! z^k}{k! [1-\alpha]_{m-k} [1-\beta]_{m-k}}$$

$$+ \sum_{n=1}^{\infty} \frac{[\alpha]_n [\beta]_n}{[m+1]_n n!} \sum_{k=0}^{n-1} \left(\frac{1}{\alpha+k} + \frac{1}{\beta+k} - \frac{1}{m+1+k} - \frac{1}{1+k} \right) z^n$$

$$\tag{12.76}$$

(there is no first term when $m = 0$ in Eq. (12.76)), then

(1) when $1 - \gamma = 0$

$$w_1(z) = F(\alpha, \beta; 1; z) \tag{12.77a}$$

$$w_2(z) = F(\alpha, \beta; 1; z) \log z + F_1(\alpha, \beta; 1; z), \tag{12.77b}$$

(2) when $1 - \gamma = -m \ (< 0)$

$$w_1(z) = F(\alpha, \beta; m+1; z) \tag{12.78a}$$

$$w_2(z) = F(\alpha, \beta; m+1; z) \log z + F_1(\alpha, \beta; m+1; z), \tag{12.78b}$$

(3) when $1 - \gamma = m \ (> 0)$

$$w_1(z) = z^m F(\alpha+m, \beta+m; m+1; z) \tag{12.79a}$$

$$w_2(z) = z^m F(\alpha+m, \beta+m; m+1; z) \log z$$

$$+ z^m F_1(\alpha+m, \beta+m; m+1; z). \tag{12.79b}$$

12.2.2.3   *Gauss hypergeometric functions II. Solution around regular singular point $z = 1$*

The convergence radius of the solution around $z = 0$, as found in Sec. 12.2.2.2, is 1. Therefore, we need to find the solution around the regular singular point $z = 1$ separately.

Letting $1 - z = \zeta$ in Eq. (12.64) to obtain the solution around the regular singular point $z = 1$, we obtain

$$\zeta(1-\zeta)\frac{d^2}{d\zeta^2}w(\zeta) + [\alpha+\beta+1-\gamma-(\alpha+\beta+1)\zeta]\frac{d}{d\zeta}w(\zeta) - \alpha\beta w(\zeta) = 0. \tag{12.80}$$

If we let $\alpha' = \alpha$, $\beta' = \beta$, $\gamma' = \alpha + \beta + 1 - \gamma$, this is a hypergeometric differential equation using these as parameters. Therefore, if γ' is not an integer, which is to say, if $\alpha + \beta - \gamma$ is not an integer, the fundamental solutions are

$$w_1(z) = F(\alpha, \beta; \alpha + \beta + 1 - \gamma; 1 - z) \tag{12.81a}$$

$$w_2(z) = (1-z)^{\gamma - \alpha - \beta} F(\gamma - \alpha, \gamma - \beta; \gamma - \alpha - \beta + 1; 1 - z). \tag{12.81b}$$

The convergence radius of this series is still 1 (taking $z = 1$ as the center). We can also follow Sec. 12.2.2.2, if $\alpha + \beta - \gamma$ is an integer.

12.2.2.4 Gauss hypergeometric functions III. Solution around regular singular point $z = \infty$

Let $z = 1/\zeta$ in Eq. (12.64) to obtain solutions around the regular singular point $z = \infty$. From this, we first obtain

$$\zeta^2(1-\zeta)^2 \frac{d^2}{d\zeta^2} w(\zeta) + \zeta(1-\zeta)[1 - \alpha - \beta - (2-\gamma)\zeta] \frac{d}{d\zeta} w(\zeta) + \alpha\beta(1-\zeta)w(\zeta) = 0.$$

The exponents at $\zeta = 0$ are α, β. Therefore, further letting $w = \zeta^\alpha u$ produces

$$\zeta(1-\zeta) \frac{d^2}{d\zeta^2} u(\zeta) + [\alpha - \beta + 1 - (2\alpha - \gamma + 2)\zeta] \frac{d}{d\zeta} u(\zeta) - \alpha(\alpha - \gamma + 1) u(\zeta) = 0,$$

and we again obtain a hypergeometric differential equation for $u(\zeta)$. Therefore, provided $\alpha - \beta$ is not an integer, we obtain

$$w_1(z) = \left(\frac{1}{z}\right)^\alpha F\left(\alpha, \alpha - \gamma + 1; \alpha - \beta + 1; \frac{1}{z}\right) \tag{12.82a}$$

$$w_2(z) = \left(\frac{1}{z}\right)^\beta F\left(\beta, \beta - \gamma + 1; \beta - \alpha + 1; \frac{1}{z}\right) \tag{12.82b}$$

as the fundamental solutions. We can also follow Sec. 12.2.2.2 if $\alpha - \beta$ is an integer.

12.2.2.5 Relations between hypergeometric functions and functions that we already know

Several complex functions that we have already seen are special cases of hypergeometric functions.

(1) Elliptic functions

$$\int_0^{\pi/2} \frac{1}{\sqrt{1 - k^2 \sin^2 \phi}} \, d\phi = \frac{\pi}{2} F\left(\frac{1}{2}, \frac{1}{2}; 1; k^2\right) \tag{12.83a}$$

$$\int_0^{\pi/2} \sqrt{1 - k^2 \sin^2 \phi} \, d\phi = \frac{\pi}{2} F\left(-\frac{1}{2}, \frac{1}{2}; 1; k^2\right) \tag{12.83b}$$

(2) Elementary functions

$$(1 + z)^n = F(-n, 1; 1; -z) \tag{12.84a}$$

$$(1 + z)^n + (1 - z)^n = 2F\left(-\frac{1}{2}n, -\frac{1}{2}n + \frac{1}{2}; \frac{1}{2}; z^2\right) \tag{12.84b}$$

$$\log(1 + z) = zF(1, 1; 2; -z) \tag{12.84c}$$

$$\log\frac{1 + z}{1 - z} = 2zF\left(\frac{1}{2}, 1, \frac{3}{2}; z^2\right). \tag{12.84d}$$

12.2.3   Broad properties of hypergeometric differential equations

Up to the preceding section, we discussed the method of finding a solution of Gauss hypergeometric differential equation as the series solution around its regular singular points. Each of them has its convergence radius, as shown in Eq. (12.73). The central theme of this section is that a solution around a given regular singular point can be expanded to a broader domain by the analytic continuation.

12.2.3.1   24 independent solutions (Kummer's transformation formula)

Use presentation by Riemann scheme (12.65), we can solve many hypergeometric differential equations. Suppose that none of $\gamma, \gamma - \alpha - \beta, \alpha - \beta$ is an integer.

Multiplying equation by $z^\delta(1 - z)^\epsilon$, the series solution around $z = 0$ changes by δ, while the series solution around $z = 1$ changes by ϵ, and the exponent of the series solution around $z = \infty$ changes by $\delta + \epsilon$. Therefore, we (expect to) obtain

$$z^\delta(1 - z)^\epsilon P \left\{ \begin{array}{ccc} 0 & 1 & \infty \\ 0 & 0 & \alpha' \ z \\ 1 - \gamma' & \gamma' - \alpha' - \beta' & \beta' \end{array} \right\}$$

$$= P \left\{ \begin{array}{ccc} 0 & 1 & \infty \\ \delta & \epsilon & \alpha' - \delta - \epsilon \ z \\ 1 - \gamma' + \delta & \gamma' - \alpha' - \beta' + \epsilon & \beta' - \delta - \epsilon \end{array} \right\}. \tag{12.85}$$

For this to satisfy the same hypergeometric differential equation as $F(\alpha, \beta; \gamma; z)$, it must equal

$$(\delta, 1 - \gamma' + \delta) = (0, 1 - \gamma) \quad \text{or} \quad (1 - \gamma, 0)$$

$$(\alpha' - \delta - \epsilon, \beta' - \delta - \epsilon) = (\alpha, \beta) \quad \text{or} \quad (\beta, \alpha)$$

$$(\epsilon, \gamma' - \alpha' - \beta' + \epsilon) = (0, \gamma - \alpha - \beta) \quad \text{or} \quad (\gamma - \alpha - \beta, 0).$$

In two different choices in the second set of above equations, they only differ by changing α for β, so $\alpha' - \delta - \epsilon = \alpha$, $\beta' - \delta - \epsilon = \beta$ may be used. Therefore, we will study the followings:

$$(\alpha' - \delta - \epsilon, \beta' - \delta - \epsilon) = (\alpha, \beta) \tag{12.86a}$$

$$(\delta, 1 - \gamma' + \delta) = (0, 1 - \gamma) \quad \text{or} \quad (1 - \gamma, 0) \tag{12.86b}$$

$$(\epsilon, \gamma' - \alpha' - \beta' + \epsilon) = (0, \gamma - \alpha - \beta) \quad \text{or} \quad (\gamma - \alpha - \beta, 0). \tag{12.86c}$$

Four solutions can be obtained, depending on how the second and third are selected.

(1) If $(\delta, 1 - \gamma' + \delta) = (0, 1 - \gamma)$, $(\epsilon, \gamma' - \alpha' - \beta' + \epsilon) = (0, \gamma - \alpha - \beta)$;

$$\delta = \epsilon = 0, \quad \alpha' = \alpha, \quad \beta' = \beta, \quad \gamma' = \gamma,$$

$$w(z) = F(\alpha, \beta; \gamma; z). \tag{12.87a}$$

(2) If $(\delta, 1 - \gamma' + \delta) = (1 - \gamma, 0)$, $(\epsilon, \gamma' - \alpha' - \beta' + \epsilon) = (\gamma - \alpha - \beta, 0)$;

$$\delta = 1 - \gamma, \quad \epsilon = \gamma - \alpha - \beta, \quad \alpha' = 1 - \beta, \quad \beta' = 1 - \alpha, \quad \gamma' = 2 - \gamma,$$

$$w(z) = z^{1-\gamma}(1 - z)^{\gamma - \alpha - \beta} F(1 - \alpha, 1 - \beta; 2 - \gamma; z). \tag{12.87b}$$

(3) If $(\delta, 1 - \gamma' + \delta) = (0, 1 - \gamma)$, $(\epsilon, \gamma' - \alpha' - \beta' + \epsilon) = (\gamma - \alpha - \beta, 0)$;

$$\delta = 0, \quad \epsilon = \gamma - \alpha - \beta, \quad \alpha' = \gamma - \beta, \quad \beta' = \gamma - \alpha, \quad \gamma' = \gamma,$$

$$w(z) = (1 - z)^{\gamma - \alpha - \beta} F(\gamma - \alpha, \gamma - \beta; \gamma; z). \tag{12.87c}$$

(4) If $(\delta, 1 - \gamma' + \delta) = (1 - \gamma, 0)$, $(\epsilon, \gamma' - \alpha' - \beta' + \epsilon) = (0, \gamma - \alpha - \beta)$;

$$\delta = 1 - \gamma, \quad \epsilon = 0, \quad \alpha' = \alpha + 1 - \gamma, \quad \beta' = \beta + 1 - \gamma, \quad \gamma' = 2 - \gamma,$$

$$w(z) = z^{1-\gamma} F(\alpha + 1 - \gamma, \beta + 1 - \gamma; 2 - \gamma; z). \tag{12.87d}$$

Furthermore, the linear transformations that transform these three regular singular points $(0, 1, \infty)$ among them are limited to the following six:

$$z' = z, \quad \frac{1}{z}, \quad 1 - z, \quad \frac{z - 1}{z}, \quad \frac{z}{z - 1}, \quad \frac{1}{1 - z}. \tag{12.88}$$

For example, let us consider

$$z^{\delta}(1 - z)^{\epsilon} F\left(\alpha', \beta'; \gamma'; \frac{z}{z - 1}\right).$$

To find the condition under which this that satisfy Gauss hypergeometric differential equation that $F(\alpha, \beta; \gamma; z)$ satisfies, we transform z to z' as $z' = z/(z - 1)$. Then this transformation moves the regular singular points $0, 1, \infty$ to $0, \infty, 1$, respectively. If we investigate in the same way as for Eqs. (12.85)–(12.87d), we obtain one solution

$$\delta = 0, \quad \epsilon = -\alpha, \quad \alpha' = \alpha, \quad \beta' = \gamma - \beta, \quad \gamma' = \gamma.$$

In this way, we can see that

$$(1-z)^{-\alpha} F\left(\alpha, \gamma - \beta; \gamma; \frac{z}{z-1}\right) \tag{12.89}$$

is one of the solutions that satisfy the same hypergeometric differential equation.

As above, we can obtain $4 \times 6 = 24$ independent solutions for all Gauss hypergeometric differential equations. This transformation is called Kummer's transformation formula. We give only the result below. (1) and (3) are the two series solutions around $z = 0$ that were described in Sec. 12.2.2.2 in this section, (5) and (6) are the two series solutions around $z = 1$ that were described in Sec. 12.2.2.3 in this section, and (11) and (12) are the two series solutions around $z = \infty$ that were described in Sec. 12.2.2.4 in this section.

- Fundamental solutions around $z = 0$: Convergence region $|z| < 1$

  (1) $F(\alpha, \beta; \gamma; z)$
  (2) $(1-z)^{\gamma-\alpha-\beta} F(\gamma - \alpha, \gamma - \beta; \gamma, z)$
  (3) $z^{1-\gamma} F(1 + \alpha - \gamma, 1 + \beta - \gamma; 2 - \gamma; z)$
  (4) $z^{1-\gamma}(1-z)^{\gamma-\alpha-\beta} F(1 - \alpha, 1 - \beta; 2 - \gamma, z)$

- Fundamental solutions around $z = 1$: Convergence region $|1 - z| < 1$

  (5) $F(\alpha, \beta; \alpha + \beta - \gamma + 1; 1 - z)$
  (6) $(1-z)^{\gamma-\alpha-\beta} F(\gamma - \alpha, \gamma - \beta; \gamma - \alpha - \beta + 1; 1 - z)$
  (7) $z^{1-\gamma} F(\alpha - \gamma + 1, \beta - \gamma + 1; \alpha + \beta - \gamma + 1; 1 - z)$
  (8) $z^{1-\gamma}(1-z)^{\gamma-\alpha-\beta} F(1 - \alpha, 1 - \beta; \gamma - \alpha - \beta + 1; 1 - z)$

- Fundamental solutions around $z = \infty$: Convergence region $|z| > 1$

  (9) $z^{-\alpha}\left(1 - \frac{1}{z}\right)^{\gamma-\alpha-\beta} F\left(1 - \beta, \gamma - \beta; 1 + \alpha - \beta; \frac{1}{z}\right)$
  (10) $z^{-\beta}\left(1 - \frac{1}{z}\right)^{\gamma-\alpha-\beta} F\left(1 - \alpha, \gamma - \alpha; 1 - \alpha + \beta; \frac{1}{z}\right)$
  (11) $z^{-\alpha} F\left(\alpha, \alpha - \gamma + 1; \alpha - \beta + 1; \frac{1}{z}\right)$
  (12) $z^{-\beta} F\left(\beta, \beta - \gamma + 1; \beta - \alpha + 1; \frac{1}{z}\right)$

- Fundamental solutions around $z = \infty$: Convergence region $|1 - z| > 1$

  (13) $(z-1)^{-\alpha} F\left(\alpha, \gamma - \beta; \alpha - \beta + 1; \frac{1}{1-z}\right)$
  (14) $(z-1)^{-\beta} F\left(\beta, \gamma - \alpha; \beta - \alpha + 1; \frac{1}{1-z}\right)$
  (15) $z^{1-\gamma}(z-1)^{\gamma-\alpha-1} F\left(\alpha - \gamma + 1, 1 - \beta; \alpha - \beta + 1; \frac{1}{1-z}\right)$

(16) $z^{1-\gamma}(z-1)^{\gamma-\beta-1}F\left(\beta-\gamma+1,1-\alpha;\beta-\alpha+1;\dfrac{1}{1-z}\right)$

- Fundamental solutions around $z = 0$: Convergence region $\operatorname{Re} z < 1/2$

(17) $(1-z)^{-\alpha}F\left(\alpha,\gamma-\beta;\gamma;\dfrac{z}{z-1}\right)$

(18) $(1-z)^{-\beta}F\left(\beta,\gamma-\alpha;\gamma;\dfrac{z}{z-1}\right)$

(19) $\left(\dfrac{z}{1-z}\right)^{1-\gamma}(1-z)^{-\alpha}F\left(\alpha-\gamma+1,1-\beta;2-\gamma;\dfrac{z}{z-1}\right)$

(20) $\left(\dfrac{z}{1-z}\right)^{1-\gamma}(1-z)^{-\beta}F\left(\beta-\gamma+1,1-\alpha;2-\gamma;\dfrac{z}{z-1}\right)$

- Fundamental solutions around $z = 1$: Convergence region $\operatorname{Re} z > 1/2$

(21) $\left(\dfrac{1-z}{z}\right)^{\gamma-\alpha-\beta}z^{-\alpha}F\left(1-\beta,\gamma-\beta;\gamma-\alpha-\beta+1;\dfrac{z-1}{z}\right)$

(22) $z^{-\alpha}F\left(\alpha,\alpha-\gamma+1;\alpha+\beta-\gamma+1;\dfrac{z-1}{z}\right)$

(23) $z^{-\beta}F\left(\beta,\beta-\gamma+1;\alpha+\beta-\gamma+1;\dfrac{z-1}{z}\right)$

(24) $\left(\dfrac{1-z}{z}\right)^{\gamma-\alpha-\beta}z^{-\beta}F\left(1-\alpha,\gamma-\alpha;\gamma-\alpha-\beta+1;\dfrac{z-1}{z}\right).$

(See Refs. [E. E. Kummer (1836)], [T. Inui (1962)], [H. Hochstadt (2012)], [M. Abramowitz and I. A. Stegun (1983)].)

12.2.3.2 *Analytic continuation and monodromy groups*

As written in Riemann scheme under the conditions of Eq. (12.60), all global information of functions can be completely determined from the local information such as the positions of regular singular points and the exponents at those points only in the case of Gauss hypergeometric functions which is a solution of the second-order ordinary differential equation with three definite singular points. In Sec. 12.2.3.1, we considered the solutions around the regular singular points $z = 0, 1, \infty$.

If $1 - \gamma$, $\alpha - \beta$, $\gamma - \alpha - \beta$ are not integers, we choose and consider a pair of fundamental solutions $(w_1^{(0)}(z),\ w_2^{(0)}(z))$ for Gauss hypergeometric differential equation Eq. (12.64), such as

$$w_1^{(0)}(z) = F(\alpha,\beta;\gamma;z)$$

$$w_2^{(0)}(z) = z^{1-\gamma}F(1+\alpha-\gamma,1+\beta-\gamma;2-\gamma;z).$$

$(w_1^{(0)},\ w_2^{(0)})$ converges in $|z| < 1$ and diverges in $|z| > 1$. The following discussion is equivalent to an analytic continuation that goes beyond the domain of $|z| < 1$, extending to a domain that does not include $z = 0, 1, \infty$.

For example, the fundamental solution that is defined in the neighborhood of $z = 1$,

$$w_1^{(1)}(z) = F(\alpha, \beta; \alpha + \beta - \gamma + 1; 1 - z),$$

$$w_2^{(1)}(z) = (1 - z)^{\gamma - \alpha - \beta} F(\gamma - \alpha, \gamma - \beta; \gamma - \alpha - \beta + 1; 1 - z)$$

converges in $|1 - z| < 1$ and diverges in $|1 - z| > 1$. From this, in the intersection domain of $|z| < 1$ and $|1 - z| < 1$, where the fundamental solution around $z = 0$ converges, $w_1^{(0)}$, $w_2^{(0)}$ should be expressed by a linear combination of $w_1^{(1)}$, $w_2^{(1)}$. This is written as

$$w_1^{(0)} = a_1 w_1^{(1)} + b_1 w_2^{(1)}, \qquad w_2^{(0)} = c_1 w_1^{(1)} + d_1 w_2^{(1)}. \tag{12.90}$$

The right-hand sides of these equations can be regarded as the analytic continuation of the domains of $w_1^{(0)}$, $w_2^{(0)}$ to $|1 - z| < 1$, $|z| > 1$.

The fundamental solution in the neighborhood of $z = \infty$

$$w_1^{(\infty)}(z) = z^{-\alpha} F\left(\alpha, \alpha - \gamma + 1; \alpha - \beta + 1; \frac{1}{z}\right)$$

$$w_2^{(\infty)}(z) = z^{-\beta} F\left(\beta, \beta - \gamma + 1; \beta - \alpha + 1; \frac{1}{z}\right)$$

converges in $|z| > 1$ and diverges in $|z| < 1$. Therefore, $w_1^{(1)}$ and $w_2^{(1)}$ are the analytic continuation of $w_1^{(\infty)}$, $w_2^{(\infty)}$ in their intersection domain, and are written as their linear combination

$$w_1^{(1)} = a_2 w_1^{(\infty)} + b_2 w_2^{(\infty)}, \qquad w_2^{(1)} = c_2 w_1^{(\infty)} + d_2 w_2^{(\infty)}. \tag{12.91}$$

Therefore, again

$$w_1^{(0)} = (a_1 a_2 + b_1 c_2) w_1^{(\infty)} + (a_1 b_2 + b_1 d_2) w_2^{(\infty)},$$

$$w_2^{(0)} = (c_1 a_2 + d_1 c_2) w_1^{(\infty)} + (c_1 b_2 + d_1 d_2) w_2^{(\infty)} \tag{12.92}$$

are the analytic continuations of $w_1^{(0)}$, $w_2^{(0)}$ in $|z| > 1$. The above is an overview of the form that results from analytic continuation for the fundamental solution in a domain including regular singular points.

In this way, we can start from $w_1^{(0)}$, $w_2^{(0)}$ and perform analytic continuation into domains not including $z = 0, 1, \infty$. We will write these as w_1, w_2. As $F(\cdot \ \cdot; z)$ is single-value holomorphic in the region of $|z| < 1$, it follows that w_1, w_2 become w_1, $\exp[2(1 - \gamma)\pi i] w_2$ when z goes around $z = 0$ in the positive direction. This factor comes from the prefactor $z^{1-\gamma}$. Thus, (w_1, w_2) undergoes the (linear) transformation represented by the matrix

$$\Omega_0 = \begin{pmatrix} 1 & 0 \\ 0 & e^{2(1-\gamma)\pi i} \end{pmatrix} \tag{12.93a}$$

as

$$\left(w_1, e^{2(1-\gamma)\pi i} w_2\right) = (w_1, w_2) \begin{pmatrix} 1 & 0 \\ 0 & e^{2(1-\gamma)\pi i} \end{pmatrix}. \qquad (12.93b)$$

Let us write the linear transformations undergone when z goes around $z = 1$ or $z = \infty$ in the positive direction as

$$\Omega_1, \qquad \Omega_\infty.$$

Going around $z = 0$ and $z = 1$ in the positive direction is the same as going around $z = \infty$ in the negative direction. Therefore, a contour going around $z = \infty$ in the positive direction after going around $z = 0$ and $z = 1$ in the positive direction makes one return back to the original point on z plane. Therefore, we get

$$\Omega_\infty = \left(\Omega_0 \Omega_1\right)^{-1}. \qquad (12.94)$$

The linear transformations Ω_0, Ω_1, Ω_∞ are called fundamental transformations and the group of these linear transformations is called the monodromy group.

Example 12.4. The following example shows how the hypergeometric function expressed by a series around $z = 0$ is connected to the hypergeometric function expressed by a series around $z = 1, \infty$ (analytic continuation). We recommend that readers interested in how the results were derived should refer to Refs. [H. Hochstadt (2012)] and [M. Abramowitz and I. A. Stegun (1983)].

$$F(\alpha, \beta; \gamma; z) = (1 - z)^{-\alpha} F\left(\alpha, \gamma - \beta; \gamma; \frac{z}{z - 1}\right) \qquad (12.95)$$

$$= \frac{\Gamma(\gamma)\Gamma(\gamma - \alpha - \beta)}{\Gamma(\gamma - \alpha)\Gamma(\gamma - \beta)} F(\alpha, \beta; 1 - \gamma + \alpha + \beta; 1 - z)$$

$$+ \frac{\Gamma(\gamma)\Gamma(\alpha + \beta - \gamma)}{\Gamma(\alpha)\Gamma(\beta)} (1 - z)^{\gamma - \alpha - \beta}$$

$$\times F(\gamma - \alpha, \gamma - \beta; 1 + \gamma - \alpha - \beta; 1 - z)$$

$$: |\arg(1 - z)| < \pi$$

$$\qquad (12.96)$$

$$= \frac{\Gamma(\gamma)\Gamma(\beta - \alpha)}{\Gamma(\gamma - \alpha)\Gamma(\beta)} (-z)^{-\alpha} F\left(\alpha, 1 + \alpha - \gamma; 1 + \alpha - \beta; \frac{1}{z}\right)$$

$$+ \frac{\Gamma(\gamma)\Gamma(\alpha - \beta)}{\Gamma(\gamma - \beta)\Gamma(\alpha)} (-z)^{-\beta} F\left(\beta, 1 + \beta - \gamma; 1 - \alpha + \beta; \frac{1}{z}\right)$$

$$: |\arg(-z)| < \pi$$

$$\qquad (12.97)$$

12.2.3.3 *Monodromy groups*

In the domain D on the complex plane, let C be a path whose start point z_0 and the end point z_1, and consider the analytic continuation of the hypergeometric function along this path C.

Analytic continuation of a holomorphic solution

$$\mathbf{w}(z) = (w_1(z), w_2(z)) \tag{12.98}$$

is performed along the path C. As we have already seen, let's assume this analytic continuation is expressed by a certain 2×2 regular matrix $\Gamma(C)$ as

$$\mathbf{w}(z)^C = \mathbf{w}(z)\,\Gamma(C). \tag{12.99}$$

If two closed paths C and C' having a common start point z_0 are mutually homotopic, we get

$$\Gamma(C) = \Gamma(C'). \tag{12.100}$$

This makes it possible to define the equivalence relationship between the paths, so a fundamental group $\pi_1(D, z_0)$ with the base point z_0 is constructed.

The $\Gamma(C)$ defined here is called a monodromy matrix of $\mathbf{w}(z)$ for C. As can be readily understood, $\Gamma(C_2 \cdot C_1) = \Gamma(C_2)\Gamma(C_1)$, $\Gamma(C^{-1}) = \Gamma(C)^{-1}$ etc. are true for various paths $C_1, C_2, \cdots$. The group of these matrices $\Gamma(C)$ is homomorphic with the fundamental group $\pi_1(D, z_0)$. This group is a monodromy group, as has already been described.

12.2.4 *Recurrence formulae of hypergeometric functions*

The hypergeometric function $F(\alpha, \beta; \gamma; z)$ satisfies the following six types of relation (recurrence formulae).

$$\left[z\frac{\mathrm{d}}{\mathrm{d}z} + \alpha\right] F(\alpha, \beta; \gamma; z) = \alpha F(\alpha + 1, \beta; \gamma; z) \tag{12.101a}$$

$$\left[z(1 - z)\frac{\mathrm{d}}{\mathrm{d}z} + (\gamma - \alpha - \beta z)\right] F(\alpha, \beta; \gamma; z) = (\gamma - \alpha)F(\alpha - 1, \beta; \gamma; z) \tag{12.101b}$$

$$\left[z\frac{\mathrm{d}}{\mathrm{d}z} + \beta\right] F(\alpha, \beta; \gamma; z) = \beta F(\alpha, \beta + 1; \gamma; z) \tag{12.101c}$$

$$\left[z(1 - z)\frac{\mathrm{d}}{\mathrm{d}z} + (\gamma - \beta - \alpha z)\right] F(\alpha, \beta; \gamma; z) = (\gamma - \beta)F(\alpha, \beta - 1; \gamma; z) \tag{12.101d}$$

$$\left[(1-z)\frac{\mathrm{d}}{\mathrm{d}z}-(\alpha+\beta-\gamma)\right]F(\alpha,\beta;\gamma;z)=\frac{(\gamma-\alpha)(\gamma-\beta)}{\gamma}F(\alpha,\beta;\gamma+1;z)$$

$$(12.101e)$$

$$\left[z\frac{\mathrm{d}}{\mathrm{d}z}+(\gamma-1)\right]F(\alpha,\beta;\gamma;z)=(\gamma-1)F(\alpha,\beta;\gamma-1;z).$$

$$(12.101f)$$

The operators on the left-hand sides of these equations, or the $(1/\alpha)[z(\mathrm{d}/\mathrm{d}z)+\alpha]$, for example, which normalizes them, are called raising or lowering operators for each α (β or γ). That is because they are operators which raise or lower α (β or γ) by one at a time.

The proof of the above equation can easily show a hypergeometric function as, for example, an equation in a specific form expanded around $z=0$. For example, using the identical relation

$$z\frac{\mathrm{d}}{\mathrm{d}z}z^k = kz^k,$$

$$\left(z\frac{\mathrm{d}}{\mathrm{d}z}+\alpha\right)\frac{[\alpha]_k[\beta]_k z^k}{k![\gamma]_k}=\frac{(\alpha+k)[\alpha]_k[\beta]_k z^k}{k![\gamma]_k}=\alpha\frac{[\alpha+1]_k[\beta]_k z^k}{k![\gamma]_k}.$$

By using these relations, Eq. (12.101a) is presented. Readers can try the same for other relations.

12.3 Differential equations which have irregular singular points: Confluent hypergeometric functions

12.3.1 *Kummer's confluent hypergeometric differential equations and confluent hypergeometric functions*

A hypergeometric differential equation has regular singular points at $0,1,\infty$. These regular singular points can be moved to any arbitrary points by a linear transformation of z (Sec. 4.3.2). Here we will consider the case in which the regular singular point 0 is left unchanged, while the regular singular points 1 and ∞ are made confluent on the one singular point ∞.[5]

Letting $z=\zeta/\beta$ in Eq. (12.64), $z=0,1,\infty$ move to $0,\beta,\infty$, respectively, and the hypergeometric differential equation becomes

$$\zeta\left(1-\frac{\zeta}{\beta}\right)\frac{\mathrm{d}^2 w(\zeta)}{\mathrm{d}\zeta^2}+\left(\gamma-\frac{\alpha+\beta+1}{\beta}\zeta\right)\frac{\mathrm{d}w(\zeta)}{\mathrm{d}\zeta}-\alpha w(\zeta)=0. \quad (12.102)$$

If we further let $\beta\to\infty$ in this situation, the regular singular point $\zeta=\beta$ becomes confluent to ∞, producing

$$\zeta\frac{\mathrm{d}^2 w(\zeta)}{\mathrm{d}\zeta^2}+(\gamma-\zeta)\frac{\mathrm{d}w(\zeta)}{\mathrm{d}\zeta}-\alpha w(\zeta)=0. \quad (12.103)$$

---

[5]The meaning of "confluence" is that two rivers merge into one.

This is called confluent hypergeometric differential equation, or Kummer's differential equation.

Through the above transformation, the regular singular point $z = 1$ approaches $z = \infty$, eventually becoming confluent. Therefore, $z = \infty$ has changed into an irregular singular point, which is a high-order singular point, rather than a regular singular point. The behavior of this solution can be obtained through a limiting operation on the solution to the hypergeometric series before confluence, and not from direct solution of Eq. (12.103). We will separately consider the cases where γ is not an integer, where $\gamma = 1$, and where $1 - \gamma = -m$ ($m = $ a positive integer).

(1) The case where γ is not an integer:
A limiting operation on Eqs. (12.72) and (12.75) produce the elementary solution

$$w_1(z) = \lim_{\beta \to \infty} F\left(\alpha, \beta; \gamma; \frac{z}{\beta}\right) = \sum_{k=0}^{\infty} \frac{[\alpha]_k}{k![\gamma]_k} z^k$$

$$\equiv F(\alpha; \gamma; z) \tag{12.104}$$

$$w_2(z) = (\text{constant}) \times \lim_{\beta \to \infty} z^{1-\gamma} F\left(\alpha + 1 - \gamma, \beta + 1 - \gamma; 2 - \gamma; \frac{z}{\beta}\right)$$

$$= z^{1-\gamma} \sum_{k=0}^{\infty} \frac{[\alpha + 1 - \gamma]_k}{k![2 - \gamma]_k} z^k$$

$$= z^{1-\gamma} F(\alpha + 1 - \gamma; 2 - \gamma; z). \tag{12.105}$$

The series of Eq. (12.104) is called confluent hypergeometric series and its convergence radius is ∞, which means it converges for all finite values of z. $F(\alpha; \gamma; z)$ is called confluent hypergeometric function.

(2) The case of $\gamma = 1$:
In this case, the first solution is just the one setting $\gamma = 1$ in Eq. (12.104). The second solution can be obtained when the regular singular point $z = 1$ is made confluent on $z = \infty$ in Eq. (12.78b).

$$w_1(z) = F(\alpha; 1; z) \tag{12.106}$$

$$w_2(z) = F(\alpha; 1; z) \log z + F_1(\alpha; 1; z), \tag{12.107}$$

where

$$F_1(\alpha; 1; z) = \lim_{\beta \to \infty} F_1\left(\alpha, \beta; 1; \frac{z}{\beta}\right) = \sum_{k=1}^{\infty} \frac{[\alpha]_k}{(k!)^2} \left[\sum_{r=0}^{k-1} \left(\frac{1}{\alpha + r} - \frac{2}{1 + r} \right) \right] z^k.$$

$$\tag{12.108}$$

(3) When $1 - \gamma = -m$, (m = positive integer):

The same method of limiting operation can be applied to Eqs. (12.78a) and (12.78b) to obtain their solutions.

$$w_1(z) = F(\alpha; m + 1; z) \tag{12.109}$$

$$w_2(z) = F(\alpha; m + 1; z) \log z + F_1(\alpha; m + 1; z). \tag{12.110}$$

Here,

$$F_1(\alpha; m + 1; z) = (-1)^{m+1} m! z^{-m} \sum_{k=0}^{m-1} \frac{(-1)^k (m - k - 1)!}{k!(\alpha - 1)(\alpha - 2) \cdots (\alpha - m + k)} z^k$$

$$+ \sum_{k=1}^{\infty} \frac{[\alpha]_k}{k![m + 1]_k} \left[\sum_{r=0}^{k-1} \left(\frac{1}{\alpha + r} - \frac{1}{m + 1 + r} - \frac{1}{1 + r} \right) \right] z^k. \tag{12.111}$$

12.3.2 Recurrence formulae of Kummer's confluent hypergeometric functions

The same kind of recurrence formulae obtained for hypergeometric functions in Sec. 12.2.4 can be obtained for confluent hypergeometric functions.

$$\left(z \frac{\mathrm{d}}{\mathrm{d}z} + \alpha \right) F(\alpha; \gamma; z) = \alpha F(\alpha + 1; \gamma; z) \tag{12.112a}$$

$$\left[z \frac{\mathrm{d}}{\mathrm{d}z} + (\gamma - \alpha - z) \right] F(\alpha; \gamma; z) = (\gamma - \alpha) F(\alpha - 1; \gamma; z) \tag{12.112b}$$

$$\left(\frac{\mathrm{d}}{\mathrm{d}z} - 1 \right) F(\alpha; \gamma; z) = \frac{\alpha - \gamma}{\gamma} F(\alpha; \gamma + 1; z) \tag{12.112c}$$

$$\left[z \frac{\mathrm{d}}{\mathrm{d}z} + (\gamma - 1) \right] F(\alpha; \gamma; z) = (\gamma - 1) F(\alpha; \gamma - 1; z). \tag{12.112d}$$

The operators on the left-hand sides of these equations, or the $(1/\alpha)[z(\mathrm{d}/\mathrm{d}z) + \alpha]$, for example, which normalizes them, are called raising or lowering operators for each α (or γ). That is because they are operators which raise or lower α (or γ) by one at a time.

These equations can be proved in a similar way as hypergeometric functions. Readers can try these proofs for themselves.

12.3.3 Solutions around irregular singular points, and asymptotic expansions

In confluent hypergeometric differential equations, the point at infinity is an irregular singular point. In this case, no solution can be expressed as a

Laurent expansion, as we have performed so far. In Eq. (12.103)

$$\frac{d^2 w(z)}{dz^2} + \left(\frac{\gamma}{z} - 1\right)\frac{dw(z)}{dz} - \frac{\alpha}{z}w(z) = 0$$

the behavior at $z \to \infty$ is determined by

$$\frac{d^2 w(z)}{dz^2} - \frac{dw(z)}{dz} = 0. \tag{12.113}$$

Therefore, substituting $w(z) = \exp \rho z$, one can obtain $\rho(\rho - 1) = 0$, then $\rho = 0, 1$. Use this ρ to search for a solution of the form $w(z) = e^{\rho z} u(z)$. We can assume a solution in the form of

$$w(z) = e^{\rho z} z^\sigma \sum_{k=0}^\infty c_k z^{-k}$$

and determine the σ and the coefficient c_k for each $\rho = 0, 1$. We will present just the results:

$$w_1(z) = z^{-\alpha} \sum_{k=0}^\infty \frac{[\alpha]_k [\alpha - \gamma + 1]_k}{k!}\left(-\frac{1}{z}\right)^k \tag{12.114}$$

$$w_2(z) = e^z z^{\alpha - \gamma} \sum_{k=0}^\infty \frac{[\gamma - \alpha]_k [1 - \alpha]_k}{k!}\left(\frac{1}{z}\right)^k. \tag{12.115}$$

These two solutions are both diverging series for all values of z. As we described asymptotic expansions and asymptotic series in the discussion of Γ functions, these series are also meaningful when we accept them as asymptotic series.

Chapter 13

Orthogonal Polynomials

Several cases in hypergeometric functions and confluent hypergeometric functions can be seen in a unified view from the perspective of orthogonal polynomials. In this chapter, by setting aside hypergeometric functions and confluent hypergeometric functions, we start by direct understanding orthogonal polynomials written with Rodrigues' formula and prepare for Chapters 14 and 15.

13.1   Orthogonal polynomials in finite interval (a, b)

13.1.1   Definition of orthogonal polynomials

Definition 13.1 (Definition of orthogonal polynomials in finite interval). Letting a, b $(a < b)$ be finite and real, consider the quadratic function $X(x)$,

$$X(x) = (x - a)(b - x) \tag{13.1}$$

and the weighting function $\rho(x)$,

$$\rho(x) = (b - x)^\alpha (x - a)^\beta \qquad (\alpha > -1,\ \beta > -1) \tag{13.2}$$

to define

$$F_n(x) = \frac{1}{\rho(x)} \frac{\mathrm{d}^n}{\mathrm{d}x^n} \left[\rho(x) X(x)^n \right]. \tag{13.3}$$

This is called **Rodrigues' formula** and the function $F_n(x)$ is a polynomial of order n for x.

If we employ Leibniz formula for differentiation, Eq. (13.3) becomes

$$F_n(x) = \sum_{k=0}^{n} \binom{n}{k} (-1)^{n-k} [\alpha + k + 1]_{n-k} [n + \beta - k + 1]_k (x - a)^{n-k} (b - x)^k,$$

$$\tag{13.4}$$

indicating that it is a polynomial of order n for x. We employ Pochhammer's symbol (12.70) here.

Definition 13.2 (Definition of Jacobi's polynomial). When $a = -1$ and $b = 1$ in Eq. (13.3), it is called Jacobi's polynomial. This is normally multiplied by a constant factor and defined as follows:

$$P_n^{(\alpha,\beta)}(x) = \frac{(-1)^n}{2^n n!}(1-x)^{-\alpha}(1+x)^{-\beta}\frac{d^n}{dx^n}[(1-x)^{n+\alpha}(1+x)^{n+\beta}]$$

$$= \frac{1}{2^n}\sum_{k=0}^{n}\binom{n+\alpha}{k}\binom{n+\beta}{n-k}(x-1)^{n-k}(x+1)^k. \qquad (13.5)$$

Definition 13.3 (Definition of Legendre's polynomial). When $\alpha = \beta = 0$ in Jacobi's polynomial (13.5), it is called Legendre's polynomial $P_n(x)$, which is very commonly seen.

$$P_n(x) = \frac{(-1)^n}{2^n n!}\frac{d^n}{dx^n}[(1-x^2)^n] = \frac{1}{2^n}\sum_{k=0}^{[n/2]}\frac{(-1)^k(2n-2k)!}{k!(n-k)!(n-2k)!}x^{n-2k}. \quad (13.6)$$

Here, $[n/2]$ expresses an integer $n/2$ when n is even, and $(n-1)/2$ when n is odd.

13.1.2   *Definition of inner product and orthogonality of orthogonal polynomial $F_n(x)$*

13.1.2.1   *Definition of inner product*

An "inner product" is defined as follows, with the weighting function $\rho(x)$.

$$(f,g) \equiv \int_a^b \rho(x)\,f(x)\,g(x)\,dx. \qquad (13.7)$$

13.1.2.2   *Orthogonality of $F_n(x)$ and $F_m(x)$ when $m \neq n$*

Under the definition of an inner product (13.7), the polynomials $F_n(x)$ and $F_m(x)$, $n \neq m$, satisfy the following orthogonal relation: Polynomials that satisfy this kind of orthogonal relation are called orthogonal polynomials.

$$\int_a^b \rho(x)\,F_n(x)\,F_m(x)\,dx = 0. \qquad (13.8)$$

Proof. We assume $n < m$ without losing the generality. If we perform a partial integration,

$$\int_a^b \rho(x)\, F_n(x)\, F_m(x)\, \mathrm{d}x = \left[F_n(x) \frac{\mathrm{d}^{m-1}}{\mathrm{d}x^{m-1}} (x-a)^{m+\beta}(b-x)^{m+\alpha} \right]_a^b$$

$$- \int_a^b \frac{\mathrm{d}F_n(x)}{\mathrm{d}x} \frac{\mathrm{d}^{m-1}}{\mathrm{d}x^{m-1}} \left[(x-a)^{m+\beta}(b-x)^{m+\alpha} \right] \mathrm{d}x$$

$$= \cdots$$

$$= (-1)^m \int_a^b \frac{\mathrm{d}^m F_n(x)}{\mathrm{d}x^m} \left[(x-a)^{m+\beta}(b-x)^{m+\alpha} \right] \mathrm{d}x = 0.$$

Here, we used the fact $\mathrm{d}^m F_n(x)/\mathrm{d}x^m = 0$ $(n < m)$. Then, the orthogonality relation between the polynomials $F_n(x)$ and $F_m(x)$ is demonstrated. $\square$

13.1.2.3 *Polynomial $F_n(x)$ of order n must be orthogonal with any arbitrary polynomial of order m $(m \leq n-1)$*

Any arbitrary polynomial $\Pi_{n-1}(x)$ of order $n-1$, defined in interval $[a, b]$, is orthogonal to $F_n(x)$.

$$\int_a^b \rho(x)\, F_n(x)\, \Pi_{n-1}(x)\, \mathrm{d}x = 0. \tag{13.9}$$

Proof. If we perform a partial integration, we get

$$\int_a^b \rho(x)\, F_n(x)\, \Pi_{n-1}(x)\, \mathrm{d}x = \left[\Pi_{n-1}(x) \frac{\mathrm{d}^{n-1}}{\mathrm{d}x^{n-1}} \rho(x)\, X(x)^n \right]_a^b$$

$$- \int_a^b \frac{\mathrm{d}\Pi_{n-1}(x)}{\mathrm{d}x} \frac{\mathrm{d}^{n-1}}{\mathrm{d}x^{n-1}} \left[\rho(x)\, X(x)^n \right] \mathrm{d}x.$$

The first term on the right-hand side certainly includes a positive power of $(x-a)(b-x)$, so is 0 at $x = a$ or $x = b$. We repeat partial integration for the second term. Carrying on in this way, we obtain the following:

$$\int_a^b \rho(x)\, F_n(x)\, \Pi_{n-1}(x)\, \mathrm{d}x = (-1)^n \int_a^b \frac{\mathrm{d}^n \Pi_{n-1}(x)}{\mathrm{d}x^n} \left[\rho(x)\, X(x)^n \right] \mathrm{d}x = 0.$$

$\square$

If we perform the above proof in the same way, it is easy to show that the following orthogonality relation is satisfied between $F_n(x)$ and any arbitrary polynomial $\Pi_m(x)$ of order $n-1$ or less $(m = 1, 2, \cdots, n-1)$.

$$\int_a^b \rho(x)\, F_n(x)\, \Pi_m(x)\, \mathrm{d}x = 0 \qquad (m = 1, 2, \cdots, n-1). \tag{13.10}$$

13.2   Orthogonal polynomials in infinite interval (a, ∞)

13.2.1   *Definition of orthogonal polynomials*

Definition 13.4 (Definition of orthogonal polynomials in infinite interval (a, ∞)).
Choosing

$$X(x) = x - a, \quad \rho(x) = e^{-x}(x - a)^{\alpha} \qquad (\alpha > -1), \tag{13.11}$$

we can define the following in Rodrigues' formula:

$$G_n(x) = \frac{1}{\rho(x)} \frac{d^n}{dx^n} \left[\rho(x) \, X(x)^n \right] \tag{13.12}$$

$G_n(x)$ is a polynomial of order n.

This can be rewritten, using Leibniz formula, as

$$G_n(x) = \sum_{k=0}^{n} \binom{n}{k} (-1)^k [\alpha + k + 1]_{n-k} (x - a)^k. \tag{13.13}$$

Equation (13.13) shows that $G_n(x)$ is a polynomial of order n for x.

Definition 13.5 (Definition of Laguerre polynomial).
When $a = 0$ and $\alpha = 0$ in Eq. (13.12), it is called Laguerre polynomial;

$$L_n(x) = e^x \frac{d^n}{dx^n} \left(e^{-x} x^n \right) = n! \sum_{k=0}^{n} (-1)^k \binom{n}{n-k} \frac{x^k}{k!}. \tag{13.14}$$

13.2.2   *Definition of inner product and orthogonality of polynomial $G_n(x)$*

An inner product is defined as follows, with the weighting function $\rho(x)$;

$$(f, g) \equiv \int_a^{\infty} \rho(x) \, f(x) \, g(x) \, dx. \tag{13.15}$$

Any arbitrary polynomial $\Pi_{n-1}(x)$ of order $n - 1$ defined in interval $[a, \infty]$, and $G_n(x)$ are orthogonal;

$$\int_a^{\infty} \rho(x) \, G_n(x) \, \Pi_{n-1}(x) \, dx = 0. \tag{13.16}$$

Proof. If we perform a partial integration in the same way as for $F_n(x)$,

$$\int_a^\infty \rho(x)\, G_n(x)\, \Pi_{n-1}(x)\, \mathrm{d}x = \left[\Pi_{n-1}(x)\frac{\mathrm{d}^{n-1}}{\mathrm{d}x^{n-1}}\mathrm{e}^{-x}(x-a)^{n+\alpha}\right]_a^\infty$$

$$- \int_a^\infty \frac{\mathrm{d}\Pi_{n-1}(x)}{\mathrm{d}x}\frac{\mathrm{d}^{n-1}}{\mathrm{d}x^{n-1}}\left[\mathrm{e}^{-x}(x-a)^{n+\alpha}\right]\mathrm{d}x$$

$$= \cdots$$

$$= (-1)^n \int_a^\infty \frac{\mathrm{d}^n \Pi_{n-1}(x)}{\mathrm{d}x^n}\left[\mathrm{e}^{-x}(x-a)^{n+\alpha}\right]\mathrm{d}x = 0.$$

$\square$

Performing the above proof in the same way, we can see that $G_n(x)$ and any arbitrary polynomial of order $n-1$ or less $\Pi_m(x)$ $(m = 1, 2, \cdots, n-1)$ are orthogonal.

The following orthogonal relation is also shown when $n \neq m$.

$$\int_a^\infty \rho(x)\, G_n(x)\, G_m(x)\, \mathrm{d}x = 0. \tag{13.17}$$

13.3 Orthogonal polynomials in infinite interval $(-\infty, \infty)$

13.3.1 *Definition of orthogonal polynomials: Hermite polynomial*

Definition 13.6 (Definition of Hermite polynomial).
Letting

$$X(x) = 1, \qquad \rho(x) = \mathrm{e}^{-x^2}, \tag{13.18}$$

we define

$$H_n(x) = (-1)^n \mathrm{e}^{x^2}\frac{\mathrm{d}^n}{\mathrm{d}x^n}\mathrm{e}^{-x^2}. \tag{13.19}$$

$H_n(x)$ is a polynomial of order n, and is called Hermite polynomial.[1]

If we use Leibniz formula for differentiation, Eq. (13.19) becomes

$$H_n(x) = \sum_{k=0}^{[n/2]} (-1)^k \frac{n!}{k!(n-2k)!}(2x)^{n-2k}. \tag{13.20}$$

Equation (13.20) tells that $H_n(x)$ is a polynomial of order n.

---

[1] We could often define the following form $\tilde{H}_n(x)$ as the Hermite polynomials; $\tilde{H}_n(x) = (-1)^n \mathrm{e}^{x^2/2}\frac{\mathrm{d}^n}{\mathrm{d}x^n}\mathrm{e}^{-x^2/2}$.

13.3.2   *Definition of inner product and orthogonality of Hermite polynomial $H_n(x)$*

An inner product is defined as follows, with the weighting function $\rho(x)$;

$$(f, g) \equiv \int_{-\infty}^{\infty} \mathrm{e}^{-x^2} f(x)\, g(x)\, \mathrm{d}x. \tag{13.21}$$

Any arbitrary polynomial $\Pi_{n-1}(x)$ of order $n - 1$ defined in interval $[-\infty, \infty]$, and $H_n(x)$ are orthogonal:

$$\int_{-\infty}^{\infty} \mathrm{e}^{-x^2} H_n(x)\, \Pi_{n-1}(x)\, \mathrm{d}x = 0. \tag{13.22}$$

Proof. If we perform a partial integration in the same way as for $F_n(x)$,

$$\int_{-\infty}^{\infty} \mathrm{e}^{-x^2} H_n(x)\, \Pi_{n-1}(x)\, \mathrm{d}x = (-1)^n \left\{ \left[\Pi_{n-1}(x) \frac{\mathrm{d}^{n-1}}{\mathrm{d}x^{n-1}} \mathrm{e}^{-x^2} \right]_{-\infty}^{\infty} \right.$$
$$\left. - \int_{-\infty}^{\infty} \frac{\mathrm{d}\Pi_{n-1}(x)}{\mathrm{d}x} \frac{\mathrm{d}^{n-1}\mathrm{e}^{-x^2}}{\mathrm{d}x^{n-1}} \mathrm{d}x \right\}$$
$$= \cdots$$
$$= (-1)^{2n} \int_{-\infty}^{\infty} \frac{\mathrm{d}^n \Pi_{n-1}(x)}{\mathrm{d}x^n} \mathrm{e}^{-x^2} \mathrm{d}x = 0.$$

$\square$

If the above proof is performed in the same way, it is also clear that $H_n(x)$ is orthogonal to any arbitrary polynomial of order $n - 1$ or less $\Pi_m(x)$ $(m = 1, 2, \cdots, n - 1)$.

The following orthogonality relation is shown where $n \neq m$;

$$\int_{-\infty}^{\infty} \mathrm{e}^{-x^2} H_n(x)\, H_m(x)\, \mathrm{d}x = 0. \tag{13.23}$$

13.4   Differential equations satisfied by orthogonal polynomials

As discussed in the present chapter, let us suppose that $\rho(x)$ is a weighting function and $X(x)$ is at most quadratic function for x. Also the orthogonal polynomial $p_n(x)$ is assumed to be defined by Rodrigues' formula

$$p_n(x) = \frac{1}{\rho(x)} \frac{\mathrm{d}^n}{\mathrm{d}x^n} [\rho(x)\, X(x)^n], \tag{13.24}$$

and to be a polynomial of order n for x. In that case, the orthogonal polynomial $p_n(x)$ satisfies the differential equation

$$X(x)\frac{\mathrm{d}^2 p_n(x)}{\mathrm{d}x^2} + p_1(x)\frac{\mathrm{d}p_n(x)}{\mathrm{d}x} + \lambda_n\, p_n(x) = 0. \tag{13.25}$$

If $p_1(x) = k_1 x + c_0$ and $X''(x) = X^{(2)}$, the constant λ_n is given by

$$\lambda_n = -nk_1 - \frac{n(n-1)}{2}X^{(2)}. \tag{13.26}$$

The proof is omitted here, but it is performed in the following two stages.

(1) $q_n(x) = X(x)p_n''(x) + p_1(x)p_n'(x) = cp_n(x)$, where c is a constant number.

(2) $c = -\lambda_n = nk_1 + \dfrac{n(n-1)}{2}X^{(2)}$.

Most of the differential equations which appear in the later chapters are categorized in this type. The differential equations that are satisfied by the orthogonal polynomials explained in this chapter are derived from Eq. (13.25), but we will only present the results here. Readers should attempt this. In the following chapters, we will see if these are Gauss hypergeometric differential equations or confluent hypergeometric differential equations.

- Legendre polynomials $P_n(x)$: Legendre's differential equations

$$(1 - x^2)\frac{\mathrm{d}^2 P_n(x)}{\mathrm{d}x^2} - 2x\frac{\mathrm{d}P_n(x)}{\mathrm{d}x} + n(n+1)P_n(x) = 0$$

- Laguerre polynomials $L_n(x)$: Laguerre's differential equations

$$x\frac{\mathrm{d}^2 L_n(x)}{\mathrm{d}x^2} + (1 - x)\frac{\mathrm{d}L_n(x)}{\mathrm{d}x} + nL_n(x) = 0$$

- Hermite polynomials $H_n(x)$: Hermite's differential equations

$$\frac{\mathrm{d}^2 H_n(x)}{\mathrm{d}x^2} - 2x\frac{\mathrm{d}H_n(x)}{\mathrm{d}x} + 2nH_n(x) = 0.$$

Chapter 14

Functions Written With Hypergeometric Functions

In this chapter, we will develop the general theory for several hypergeometric functions that appearing in physics and engineering, and summarize the necessary matters in the form of physics and engineering.

14.1 Problems in physics: physical phenomena and partial differential equations in spherically-symmetric systems

Let us examine some typical examples of hypergeometric differential equations.

14.1.1 Laplace operator by three-dimensional polar coordinates

When solutions of partial differential equations are discussed under various boundary conditions, it is often convenient to employ orthogonal curvilinear coordinate systems that are suitable for the boundary conditions.

The three-dimensional Laplace operator (Laplacian) appears in the differentiation of the spatial part in most cases of most partial differential equations in physic in three-dimensional space.

$$\Delta = \frac{\partial^2}{\partial x^2} + \frac{\partial^2}{\partial y^2} + \frac{\partial^2}{\partial z^2}. \tag{14.1}$$

When dealing with this differential operator, the discussion can be advanced using x, y, and z unchanged, if a cubic domain is considered. If, on the other hand, the boundary is a sphere, it is much more convenient to use three-dimensional polar coordinates.

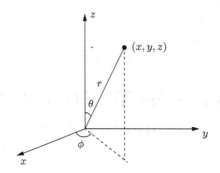

Fig. 14.1   Three-dimensional polar coordinates.

Let us represent the three-dimensional Laplace operator using the polar coordinates (r, θ, ϕ) shown in Fig. 14.1. Perform variable transformation

$$x = r \sin\theta \cos\phi, \qquad y = r \sin\theta \sin\phi, \qquad z = r \cos\theta. \tag{14.2}$$

The domain θ, ϕ is

$$0 \le \theta \le \pi, \qquad 0 \le \phi \le 2\pi.$$

Through these variable transformations, a somewhat tiresome calculation but not a difficult one, we can write the Laplace operator in three-dimensional polar coordinates as

$$\triangle = \frac{1}{r^2} \frac{\partial}{\partial r}\left(r^2 \frac{\partial}{\partial r}\right) + \frac{1}{r^2}\left[\frac{1}{\sin\theta} \frac{\partial}{\partial \theta}\left(\sin\theta \frac{\partial}{\partial \theta}\right) + \frac{1}{\sin^2\theta} \frac{\partial^2}{\partial \phi^2}\right]. \tag{14.3}$$

Consider Laplace equations in three-dimensional space shown below;

$$\left(\frac{\partial^2}{\partial x^2} + \frac{\partial^2}{\partial y^2} + \frac{\partial^2}{\partial z^2}\right)u = 0. \tag{14.4}$$

Laplace equation can be written in polar coordinates as

$$\left(\frac{\partial^2}{\partial r^2} + \frac{2}{r}\frac{\partial}{\partial r}\right)u + \frac{1}{r^2}\left[\frac{1}{\sin\theta}\frac{\partial}{\partial \theta}\left(\sin\theta \frac{\partial}{\partial \theta}\right) + \frac{1}{\sin^2\theta}\frac{\partial^2}{\partial \phi^2}\right]u = 0. \tag{14.5}$$

Assuming a separable form of a solution

$$u(r, \theta, \phi, t) = R(r)\Theta(\theta)\Phi(\phi), \tag{14.6}$$

we can rewrite the equation as

$$\frac{1}{R}\left(\frac{d^2}{dr^2} + \frac{2}{r}\frac{d}{dr}\right)R + \frac{1}{r^2}\left[\frac{1}{\Theta \sin\theta}\frac{d}{d\theta}\left(\sin\theta \frac{d\Theta}{d\theta}\right) + \frac{1}{\sin^2\theta}\frac{1}{\Phi}\frac{d^2\Phi}{d\phi^2}\right] = 0.$$

The first term and the part of $[\cdots]$ in the second term on the left-hand side are functions of only r and (θ, ϕ), respectively, and the part $[\cdots]$ must be a constant. We can discuss θ and ϕ in the term $[\cdots]$ in the same way. Rearranging them on this basis produces

$$\left(\frac{d^2}{dr^2} + \frac{2}{r}\frac{d}{dr}\right)R - \frac{\mu}{r^2}R = 0 \tag{14.7}$$

$$\frac{d^2\Phi}{d\phi^2} + \kappa\Phi = 0 \tag{14.8}$$

$$\frac{1}{\sin\theta}\frac{d}{d\theta}\left(\sin\theta\frac{d\Theta}{d\theta}\right) + \left(\mu - \frac{\kappa}{\sin^2\theta}\right)\Theta = 0 \tag{14.9}$$

where κ, μ are constants. The range of θ and ϕ is

$$0 \le \theta \le \pi, \qquad 0 \le \phi \le 2\pi \tag{14.10}$$

and $\phi = 0$ and 2π are connected, so Φ must satisfy a periodic boundary condition

$$\Phi(0) = \Phi(2\pi). \tag{14.11}$$

14.1.2   *Spherical harmonics*

We will now solve some examples of the differential equations appearing in Sec. 14.1.1. First, let us consider solutions of the type

$$R(r) = r^n, \quad (n = 0, 1, 2, \cdots). \tag{14.12}$$

Substituting that into Eq. (14.7) μ should be

$$\mu = n(n+1). \tag{14.13}$$

The periodic boundary condition (14.11) for $\Phi(\phi)$ determines

$$\Phi(\phi) = e^{im\phi} \tag{14.14}$$

$$\kappa = m^2 \quad (m = 0, \pm 1, \pm 2, \cdots). \tag{14.15}$$

14.1.2.1   *Differential equations which $\Theta(\theta)$ should satisfy*

With Eqs. (14.13) and (14.15), the differential equation (14.9) for Θ becomes

$$\frac{1}{\sin\theta}\frac{d}{d\theta}\left(\sin\theta\frac{d\Theta}{d\theta}\right) + \left[n(n+1) - \frac{m^2}{\sin^2\theta}\right]\Theta = 0. \tag{14.16}$$

In this differential equation, applying variable transformation for $\cos\theta = \omega$, writing $\Theta(\theta) = P(\omega)$ converts Eq. (14.16) into

$$\frac{d}{d\omega}\left[(1 - \omega^2)\frac{dP}{d\omega}\right] + \left[n(n+1) - \frac{m^2}{1 - \omega^2}\right]P = 0, \quad (-1 \le \omega \le 1). \tag{14.17}$$

14.1.2.2   *Solution when m = 0: Legendre polynomial*

Consider Eq. (14.17) when $m = 0$

$$(1 - \omega^2)\frac{\mathrm{d}^2}{\mathrm{d}\omega^2}P - 2\omega\frac{\mathrm{d}}{\mathrm{d}\omega}P + n(n+1)P = 0. \tag{14.18}$$

This differential equation is the one that appeared in Eq. (12.5) (Legendre differential equation). The solution to this differential equation is known to be Legendre polynomial of order n and $\omega = \pm 1$ is a regular singular point. The transformation $\omega = \frac{1}{\zeta}$ makes the differential equation (14.18) into

$$\frac{\mathrm{d}^2}{\mathrm{d}\zeta^2}P + \frac{2\zeta}{\zeta^2 - 1}\frac{\mathrm{d}}{\mathrm{d}\zeta}P + \frac{n(n+1)}{\zeta^2(\zeta^2 - 1)}P = 0,$$

so $\zeta = 0$, correspondingly $\omega = \infty$, is also a regular singular point of this differential equation.

Furthermore, performing the variable transformation with $\omega = 1 - 2\zeta$ changes Legendre differential equation (14.18) into

$$\zeta(1 - \zeta)\frac{\mathrm{d}^2 P}{\mathrm{d}\zeta^2} + (1 - 2\zeta)\frac{\mathrm{d}P}{\mathrm{d}\zeta} + n(n+1)P = 0. \tag{14.19}$$

This is the one of letting $\alpha = n + 1$, $\beta = -n$, $\gamma = 1$ in Gauss differential equation (12.64), so the first solution of the differential equation, using the notation of Gauss hypergeometric function, is

$$P_n(\omega) = F\left(-n, n + 1; 1; \frac{1 - \omega}{2}\right). \tag{14.20}$$

This is the form when Legendre polynomial of order n is seen as a hypergeometric function. This is called Legendre polynomial of the first kind. The second solution is provided by Eq. (12.77b).

14.1.2.3   *Solution when m =integer (m ≠ 0): Associated Legendre function*

Let us examine the differential equation (14.17)

$$\frac{\mathrm{d}^2}{\mathrm{d}\omega^2}P - \frac{2\omega}{1 - \omega^2}\frac{\mathrm{d}}{\mathrm{d}\omega}P + \left[\frac{n(n+1)}{1 - \omega^2} - \frac{m^2}{(\omega^2 - 1)^2}\right]P = 0.$$

Considering $-1 < \omega < 1$ here, and letting

$$P(\omega) = (1 - \omega^2)^{m/2}y(\omega), \tag{14.21}$$

the differential equation for $y(\omega)$ is

$$\frac{\mathrm{d}^2}{\mathrm{d}\omega^2}y(\omega) - 2(m + 1)\frac{\omega}{1 - \omega^2}\frac{\mathrm{d}}{\mathrm{d}\omega}y(\omega) + \left[\frac{n(n+1)}{1 - \omega^2} - \frac{m(m+1)}{1 - \omega^2}\right]y(\omega) = 0.$$

Furthermore, with $\omega = 1 - 2\zeta$, we get

$$\zeta(1 - \zeta)\frac{d^2}{d\zeta^2}y + (m + 1)(1 - 2\zeta)\frac{d}{d\zeta}y + (n - m)(n + m + 1)y = 0. \quad (14.22)$$

This is Gauss hypergeometric differential equation (12.64), so the solution is a hypergeometric function $F(m-n, n+m+1; m+1; (1-\omega)/2)$. Therefore, the first solution to Eq. (14.17) is

$$P_n^m(\omega) = \frac{\Gamma(n + m + 1)}{2^m m! \Gamma(n - m + 1)}$$

$$\times (1 - \omega^2)^{m/2} F\left(m - n, n + m + 1; m + 1; \frac{1 - \omega}{2}\right), \quad (14.23)$$

where m is an integer, either positive or negative. The coefficients in Eq. (14.23) are determined for its normalization. Here, it should be noted that ω is real in the range $-1 < \omega < 1$. In the case of a real number of $|\omega| > 1$, and the case of a complex number, it is usual to change the phase factor for the definition [defined in Eq. (14.50)], so care is required.

The second solution is discussed in Sec. 14.2 and later. $P_n^m(\omega)$ is called associated Legendre function and the differential equation (14.17) is called associated Legendre differential equation.

From the definition, one can prove the relation;

$$P_n^{-m}(\omega) = (-1)^m \frac{\Gamma(n - m + 1)}{\Gamma(n + m + 1)} P_n^m(\omega). \quad (14.24)$$

14.1.2.4 *Spherical harmonics $Y_{n,m}(\theta, \phi)$*

Considering a solution $w(r, \theta, \phi)$ of three-dimensional Laplace equation $\triangle w = 0$ and assume the form

$$w(r, \theta, \phi) = r^n Y(\theta, \phi)$$

in which $Y(\theta, \phi)$ is

$$Y_{nm}(\theta, \phi) = (-1)^{(m+|m|)/2} \sqrt{\frac{2n + 1}{4\pi} \cdot \frac{(n - |m|)!}{(n + |m|)!}} P_n^m(\cos\theta)\, e^{im\phi}. \quad (14.25)$$

This $Y_{nm}(\theta, \phi)$ is called spherical harmonics and can be written as the product of a Legendre associated function and an exponential function. The constant is determined so that the normalization integral is 1.[1]

$$\int_0^{2\pi} d\phi \int_0^{\pi} d\theta\, \overline{Y_{nm}(\theta, \phi)} Y_{n'm'}(\theta, \phi) = \delta_{n,n'} \delta_{m,m'}. \quad (14.26)$$

---

[1]There are other definitions with some different constant factors without changing the following normalization integral. One should be careful about the definitions of spherical harmonics, when reads other textbooks.

14.2 Legendre differential equation and Legendre's function: Example of P function

The previous section described Legendre's polynomials and associated Legendre functions. This section will discuss them in somewhat more general terms.

14.2.1 *Legendre differential equations*

The differential equation with an arbitrary complex number ν instead of n in Eq. (14.18)

$$(1 - z^2)\frac{d^2 w(z)}{dz^2} - 2z\frac{dw(z)}{dz} + \nu(\nu + 1)w(z) = 0 \qquad (14.27)$$

is Legendre differential equation of order ν. This differential equation is Gauss hypergeometric differential equation (12.64), and is the case of $\alpha = \nu + 1$, $\beta = -\nu$, $\gamma = 1$. Furthermore, as seen in Eq. (14.18), taking $z = 1, -1, \infty$ as regular singular points, their indices are

$$
\begin{array}{ll}
z = 1 : & 0 \quad 0 \\
z = -1 : & 0 \quad 0 \\
z = \infty : & -\nu \ \ \nu + 1
\end{array}
$$

Therefore, the solution using Riemann scheme can be expressed as

$$P(z) = P\left\{\begin{array}{ccc} 1 & -1 & \infty \\ 0 & 0 & \nu + 1 \\ 0 & 0 & -\nu \end{array} z \right\}, \qquad (14.28)$$

which is Riemann P function. Alternatively, as we have already done, letting

$$\zeta = \frac{1 - z}{2}$$

makes Legendre differential equation (14.27) into

$$\zeta(1 - \zeta)\frac{d^2 w(\zeta)}{d\zeta^2} + (1 - 2\zeta)\frac{dw(\zeta)}{d\zeta} + \nu(\nu + 1)w(\zeta) = 0. \qquad (14.29)$$

This is Gauss differential equation (12.64), taking $\alpha = \nu + 1$, $\beta = -\nu$, $\gamma = 1$. In Eq. (14.29), the position of the regular singular point shifts to $\zeta = 0, 1, \infty$, so if it is represented using Riemann scheme, the solution becomes

$$P(z) = P\left\{\begin{array}{ccc} 0 & 1 & \infty \\ 0 & 0 & \nu + 1 \\ 0 & 0 & -\nu \end{array} \frac{1 - z}{2} \right\}. \qquad (14.30)$$

14.2.2 *Legendré functions*

14.2.2.1 *Legendre functions of the first kind*

The first solution of Legendre differential equation is, as given in Eqs. (12.77a) and (12.71),

$$P_\nu(z) = F\left(\nu+1, -\nu; 1; \frac{1-z}{2}\right) = \sum_{k=0}^{\infty} \frac{[\nu+1]_k[-\nu]_k}{(k!)^2}\left(\frac{1-z}{2}\right)^k. \quad (14.31)$$

This is called Legendre function of the first kind. Figure 14.2 shows the behavior when ν changes between 0 and 4. When $\nu = n$ (integer), the sum terminates at a finite term, and it is just Legendre polynomial $P_n(z)$, which we have already seen.

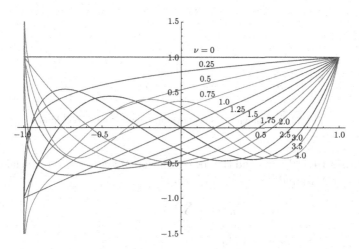

Fig. 14.2   Behavior of Legendre function of the first kind.

14.2.2.2 *Legendre functions of the second kind*

For the second solution, in a general case where ν is neither 0 nor an integer,

$$w_2(z) = F\left(\nu+1, -\nu; 1; \frac{1-z}{2}\right)\log\frac{1-z}{2} + F_1\left(\nu+1, -\nu; 1; \frac{1-z}{2}\right) \quad (14.32)$$

as given as Eq. (12.77b). However, the form commonly used is

$$Q_\nu(z) = B\left(\frac{1}{2}, \nu+1\right)\frac{1}{(2z)^{\nu+1}}F\left(\frac{\nu+1}{2}, \frac{\nu}{2}+1; \nu+\frac{3}{2}; \frac{1}{z^2}\right). \quad (14.33)$$

The function (14.33) is called Legendre function of the second kind. Figure 14.3 shows its behavior when ν changes between 0 and 4. It can be shown directly that Eq. (14.33) is the solution of Legendre differential equation around $z = \infty$.

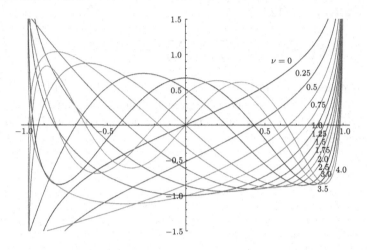

Fig. 14.3   Behavior of Legendre function of the second kind.

Example 14.1. Let us find the series solution of Legendre differential equation (14.27) around $z = \infty$. If we assume

$$w(z) = \sum_{k=0}^{\infty} c_k z^{-\rho-k},$$

$\rho = -\nu,\ \nu + 1$, and we obtain

$$c_1 = 0, \qquad (\rho + k + \nu)(\rho + k - \nu - 1)c_k - (\rho + k - 2)(\rho + k - 1)c_{k-2} = 0.$$

Letting $c_0 = 1$, we can obtain from this the first solution corresponding to $\rho = -\nu$,

$$w_1(z) = z^{\nu}\left[1 + \sum_{k=1}^{\infty} \frac{(-1)^k \nu(\nu - 1)\cdots(\nu - 2k + 1)}{2 \cdot 4 \cdots 2k \cdot (2\nu - 1)(2\nu - 3)\cdots(2\nu - 2k + 1)}z^{-2k}\right]$$

$$(14.34)$$

and the second solution corresponding to $\rho = \nu + 1$.

$$w_2(z) = z^{-\nu-1}\left[1 + \sum_{k=1}^{\infty} \frac{(\nu + 1)(\nu + 2)\cdots(\nu + 2k - 1)(\nu + 2k)}{2 \cdot 4 \cdots 2k \cdot (2\nu + 3)(2\nu + 5)\cdots(2\nu + 2k + 1)}z^{-2k}\right].$$

$$(14.35)$$

Multiplying the $w_2(z)$ found here by a constant, we obtain Eq. (14.33).

14.2.3 Recurrence formulae of Legendre functions

Now, let us consider the recurrence formula of Legendre function P_ν or Q_ν. The equations below are written for P_ν, but that can be replaced with Q_ν.

$$(z^2 - 1)\frac{\mathrm{d}P_\nu(z)}{\mathrm{d}z} = -(\nu + 1)(zP_\nu(z) - P_{\nu+1}(z)) \tag{14.36a}$$

$$(z^2 - 1)\frac{\mathrm{d}P_\nu(z)}{\mathrm{d}z} = \nu(zP_\nu(z) - P_{\nu-1}(z)) \tag{14.36b}$$

$$(z^2 - 1)\frac{\mathrm{d}P_\nu(z)}{\mathrm{d}z} = \frac{\nu(\nu + 1)}{2\nu + 1}(P_{\nu+1}(z) - P_{\nu-1}(z)) \tag{14.36c}$$

$$(\nu + 1)P_{\nu+1}(z) - (2\nu + 1)zP_\nu(z) + \nu P_{\nu-1}(z) = 0. \tag{14.36d}$$

Proof. We will describe Eq. (14.36a) in detail and only give a brief outline of the rest. Readers should attempt this by themselves.

- Proof of equation (14.36a): From Eq. (14.31), it follows from $\zeta = (1 - z)/2$ that $P_\nu(z) = F(\nu + 1, -\nu; 1, \zeta)$. Taking $\alpha = \nu + 1$, $\beta = -\nu$, $\gamma = 1$ in Eq. (12.101d) produces

$$\left[\zeta(1-\zeta)\frac{\mathrm{d}}{\mathrm{d}\zeta} + (1+\nu - (\nu+1)\zeta)\right]F(\nu+1, -\nu; 1; \zeta) = (1+\nu)F(\nu+1, -\nu-1; 1; \zeta).$$

Taking $\alpha = \nu + 1$, $\beta = -\nu - 1$, $\gamma = 1$ in Eq. (12.101a) produces

$$\left[\zeta\frac{\mathrm{d}}{\mathrm{d}\zeta} + (\nu+1)\right]F(\nu+1, -\nu-1; 1; \zeta) = (\nu+1)F(\nu+1+1, -\nu-1; 1; \zeta),$$

so

$$\left[\zeta\frac{\mathrm{d}}{\mathrm{d}\zeta} + (\nu+1)\right]\left[\zeta(1-\zeta)\frac{\mathrm{d}}{\mathrm{d}\zeta} + (\nu+1)(1-\zeta)\right]F(\nu+1, -\nu; 1; \zeta)$$
$$= (\nu+1)^2 F(\nu+2, -\nu-1; 1; \zeta).$$

If we use Eq. (14.31), it is rewritten as

$$\left[\zeta\frac{\mathrm{d}}{\mathrm{d}\zeta} + (\nu+1)\right]\left[\zeta(1-\zeta)\frac{\mathrm{d}}{\mathrm{d}\zeta} + (\nu+1)(1-\zeta)\right]P_\nu(z) = (\nu+1)^2 P_{\nu+1}(z).$$

If we retransform ζ on the left-hand side to z and, using Eq. (14.27), eliminate terms of the second derivative for z, we obtain

$$\left[(\nu+1)(z^2-1)\frac{\mathrm{d}}{\mathrm{d}z} + z(\nu+1)^2\right]P_\nu(z) = (\nu+1)^2 P_{\nu+1}(z),$$

so with further reorganization we obtain Eq. (14.36a).

For $Q_\nu(z)$, we set $\zeta = 1/z^2$ and go through the same procedure.

- Proof of Eq. (14.36b): In the same way as for Eq. (14.36a), an expression using the recurrence formula $F(\alpha, \beta; \gamma; \zeta)$ can be used. Readers should practice this by themselves.
- Proof of equation (14.36c): Eliminate $P_\nu(z)$ from Eqs. (14.36a) and (14.36b).
- Proof of Eq. (14.36d): Eliminate $dP_\nu(z)/dz$ from Eqs. (14.36a) and (14.36b).

$\square$

Example 14.2. A number of recurrence formulae can also be obtained from Eqs. (14.36a)–(14.36d). The followings are some examples. It remains valid if P_ν is replaced with Q_ν.

$$\frac{dP_{\nu+1}(z)}{dz} - z\frac{dP_\nu(z)}{dz} = (\nu + 1)P_\nu(z) \tag{14.37a}$$

$$z\frac{dP_\nu(z)}{dz} - \frac{dP_{\nu-1}(z)}{dz} = \nu P_\nu(z). \tag{14.37b}$$

Proof. We give an outline of the proof.

- Proof of Eq. (14.37a):
  From $\{(-z) \times$ Eq. (14.36a) $+ [$Eq. (14.36b)$]_{\nu \to \nu+1}\}/(z^2 - 1)$.
- Proof of Eq. (14.37b):
  From $\{[-$Eq. (14.36a)$]_{\nu \to \nu-1} + z \times$ Eq. (14.36b)$\}/(z^2 - 1)$.

$\square$

14.3 Legendre polynomials

Until now, we have considered Legendre functions of order ν. We will now discuss cases in which $\nu = n$ (0 or an integer).

14.3.1 *Legendre polynomials of the first kind*

If the order number ν of Legendre function is 0 or an integer, it becomes a polynomial in which the coefficient of the infinite series terminates at a finite value. This is a Legendre polynomial that has already been explained in Eq. (13.6).

$$\Big[P_\nu(z)\Big]_{\nu=n} = P_n(z) = \frac{1}{2^n} \sum_{k=0}^{[n/2]} \frac{(-1)^k(2n-2k)!}{k!(n-k)!(n-2k)!} x^{n-2k}.$$

Figure 14.4 shows a number of Legendre polynomials $P_n(x)$.

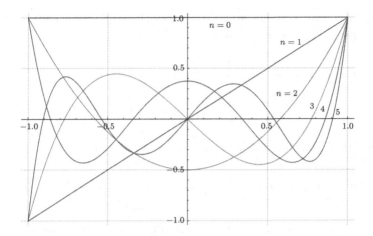

Fig. 14.4   Behavior of Legendre polynomials of the first kind.

Example 14.3. Readers should show directly that $\Big[P_\nu(z)\Big]_{\nu=n} = P_n(z)$.

14.3.2   *Legendre polynomials of the second kind*

Legendre function of the second kind is defined in Eq. (14.33). Let us set ν to be 0 or a positive integer n. The results are

$$Q_n(z) = \frac{1}{2}P_n(z)\log\frac{z+1}{z-1} - W_{n-1}(z) \tag{14.38a}$$

$$W_{n-1}(z) = \sum_{k=1}^{[(n-1)/2]} \frac{2n-4k-1}{(2k+1)(n-k)}P_{n-2k-1}(z). \tag{14.38b}$$

Here is the proof.

Proof. Employ mathematical induction for the proof. First, show that the equation holds for $n = 0$. Then if it holds for $n \le m$, and then it also holds for $n = m+1$.

When $\nu = 0$, from Eq. (14.33)

$$Q_0(z) = B\Big(\frac{1}{2}, 1\Big)\frac{1}{2z}F\Big(\frac{1}{2}, 1; \frac{3}{2}; \frac{1}{z^2}\Big)$$

$$= \frac{\Gamma(\frac{1}{2})\Gamma(1)}{\Gamma(\frac{3}{2})}\frac{1}{2z}\sum_{k=0}^{\infty}\frac{[\frac{1}{2}]_k[1]_k}{[\frac{3}{2}]_k k!}\frac{1}{z^{2k}} = \frac{1}{z}\sum_{k=0}^{\infty}\frac{1}{2k+1}\cdot\frac{1}{z^{2k}}$$

$$= \frac{1}{2}\Big[\log\Big(1+\frac{1}{z}\Big) - \log\Big(1-\frac{1}{z}\Big)\Big] = \frac{1}{2}\log\frac{z+1}{z-1} \tag{14.39}$$

and Eq. (14.38a) is true.

Assume that it is true when ν is a positive integer such that $\nu \leq m$. Equation (14.32) is true, so for $k = 0, 1, 2, \cdots, m$,

$$
\begin{aligned}
Q_{k+1}(z) &= \left[\frac{z^2 - 1}{k + 1} \cdot \frac{\mathrm{d}}{\mathrm{d}z} + z \right] Q_k(z) \\
&= \left[\frac{z^2 - 1}{k + 1} \cdot \frac{\mathrm{d}}{\mathrm{d}z} + z \right] \left(\frac{1}{2} P_k(z) \log \frac{z + 1}{z - 1} - W_{k-1}(z) \right)
\end{aligned}
$$

is true. If we calculate this and use Eq. (14.36a), we find

$$
Q_{k+1}(z) = \frac{1}{2} P_{k+1}(z) \log \frac{z + 1}{z - 1} - \frac{1}{k + 1} P_k(z) - \left[\frac{z^2 - 1}{k + 1} \cdot \frac{\mathrm{d}}{\mathrm{d}z} + z \right] W_{k-1}(z).
$$

Furthermore, using Eq. (14.36a) repeatedly in the calculation of $\left[\frac{z^2 - 1}{k + 1} \cdot \frac{\mathrm{d}}{\mathrm{d}z} + z \right] W_{k-1}(z)$ demonstrates

$$
\frac{1}{k + 1} P_{k+1}(z) + \left[\frac{z^2 - 1}{k + 1} \cdot \frac{\mathrm{d}}{\mathrm{d}z} + z \right] W_{k-1}(z) = W_k(z).
$$

The above proves that Eq. (14.38a) is true when $n = m + 1$.   □

Figure 14.5 shows several Legendre polynomials of the second kind $Q_n(x)$.

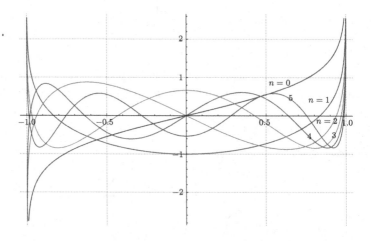

Fig. 14.5 Behavior of Legendre polynomials of the second kind.

14.3.3 Orthogonal relations and normalization integrals of Legendre polynomials

Legendre polynomial satisfies the following orthogonal relation.

$$\int_{-1}^{1} P_n(x) \, P_m(x) \, dx = 0 \qquad (n \neq m). \qquad (14.40)$$

We have already proved this in general with Eq. (13.8) in Sec. 13.1.2.2, so we need not repeat it here.

The normalization integral of Legendre polynomial of order n is

$$\int_{-1}^{1} P_n(x)^2 dx = \frac{2}{2n+1}. \qquad (14.41)$$

Use Rodrigues' formula (13.6) to derive the normalization integral (14.41).

Proof. From Eq. (13.6) (using a sequence of partial integrations)

$$
\begin{aligned}
\int_{-1}^{1} P_n(x)^2 dx &= \frac{1}{(2^n n!)^2} \int_{-1}^{1} \left[\frac{d^n}{dx^n} (1-x^2)^n \right]^2 dx \\
&= \frac{(-1)^n}{(2^n n!)^2} \int_{-1}^{1} (x^2-1)^n \frac{d^{2n}}{dx^{2n}} (x^2-1)^n dx \\
&= \frac{(-1)^n (2n)!}{(2^n n!)^2} \int_{-1}^{1} (x^2-1)^n dx \\
&= \frac{2}{2n+1}
\end{aligned}
$$

and Eq. (14.41) is derived. □

14.3.4 Generating function of Legendre polynomial

Legendre polynomial $P_n(x)$ satisfies the equation

$$\frac{1}{\sqrt{1 - 2xt + t^2}} = \sum_{n=0}^{\infty} P_n(x) \, t^n. \qquad (14.42)$$

The right-hand side is called the generating function expansion of Legendre polynomial, and the left-hand side is called the generating function.

Proof of Eq. (14.42). From Eq. (13.6) according to Goursat's theorem,

$$P_n(x) = \frac{1}{2\pi i} \oint_C \frac{(z^2-1)^n}{2^n (z-x)^{n+1}} dz. \qquad (14.43)$$

The contour path C is any arbitrary closed curve whose center is $z = 0$, includes $z = x$ inside it, and goes around in a positive direction. Let this

closed curve be a circle of radius 1. (For now, the definition is $1 \geq x \geq -1$.)
The integrated function has $z = x$ as a pole of order $n + 1$.

Here, letting

$$R = \sqrt{1 - 2xt + t^2},$$

we use

$$z = \frac{1 - R}{t} \tag{14.44}$$

to transform the variable from z to t and obtain

$$dz = -\frac{dt}{t^2}(1 - R) + \frac{1}{t}\frac{x - t}{R}dt = \frac{1 - R - xt}{t^2 R}dt$$

$$\frac{1}{z - x} = \frac{t}{1 - R - xt}, \quad z^2 - 1 = \frac{2}{t^2}(1 - R - xt).$$

The contour path[2] that goes around $z = x$ in the positive direction is
changed into the contour path C' that goes around $t = 0$ in the positive
direction on the t plane. We obtain

$$P_n(x) = \frac{1}{2\pi i} \oint_{C'} \left(\frac{1}{t}\right)^{n+1}\frac{dt}{R} = \frac{1}{n!}\left\{\frac{d^n}{dt^n}\left(\frac{1}{R}\right)\right\}_{t=0}. \tag{14.45}$$

From this, we get

$$\sum_{n=0}^{\infty} P_n(x)\, t^n = \sum_{n=0}^{\infty} \frac{1}{n!}\left\{\frac{d^n}{dt^n}\left(\frac{1}{R}\right)\right\}_{t=0} t^n. \tag{14.46}$$

The right-hand side of Eq. (14.46) can only be the Taylor series expansion
of $1/R$ around $t = 0$. Thus, the right-hand side is $1/R$ itself, and we obtain
Eq. (14.42). $\qquad\square$

Example 14.4. The proof here for Eq. (14.44) appears somewhat *a priori*.
It is recognized as correct, of course, but difficult to reach by oneself. Let
us show this directly.

To show Eq. (14.42), we should calculate

$$\sum_{n=0}^{\infty} P_n(x)\, t^n = \sum_{n=0}^{\infty} \frac{1}{2\pi i} \oint_{|z|=1} \frac{t^n(z^2 - 1)^n}{2^n(z - x)^{n+1}}dz.$$

Using the fact that in the range $\left|\frac{t(z^2 - 1)}{2(z - x)}\right| < 1$, it can be arranged as

$$\sum_{n=0}^{\infty} \frac{t^n(z^2 - 1)^n}{2^n(z - x)^n} = \frac{2(z - x)}{-tz^2 + 2z - 2x + t}$$

---

[2] $\frac{2(z - x)}{z^2 - 1} = t$, so if z moves around x in a sufficiently small circle in the direction that
increases the argument, t moves around 0 in the direction that increases the argument.

we find

$$\sum_{n=0}^{\infty} P_n(x)\, t^n = \frac{1}{2\pi i} \oint_{|z|=1} \frac{2}{-tz^2 + 2z - 2x + t}\, dz.$$

Equation (14.42) is obtained by using residue calculation to find the integration of the right-hand side.

14.4 Associated Legendre functions and associated Legendre polynomials

14.4.1 *Associated Legendre's differential equations and associated Legendre functions*

We already know a differential equation (14.17) that associated Legendre polynomial satisfies. Here, we take $n \to \nu$, $m \to \mu$ and generalize each to a complex number.

$$\frac{d}{dz}\left[(1 - z^2)\frac{dw(z)}{dz}\right] + \left[\nu(\nu + 1) - \frac{\mu^2}{1 - z^2}\right]w(z) = 0. \tag{14.47}$$

This is associated Legendre's differential equation. This is Fuchsian differential equation with $z = \pm 1, \infty$ as a regular singular point, and its solution can be expressed with Riemann scheme as

$$w(z) = P\left\{\begin{array}{ccc} 1 & -1 & \infty \\ \mu/2 & \mu/2 & \nu + 1 \ z \\ -\mu/2 & -\mu/2 & -\nu \end{array}\right\} \tag{14.48a}$$

$$= (1 - z^2)^{\mu/2} P\left\{\begin{array}{ccc} 1 & -1 & \infty \\ 0 & 0 & \nu + \mu + 1 \ z \\ -\mu & -\mu & -\nu + \mu \end{array}\right\}. \tag{14.48b}$$

This is the case because if we multiply Eq. (14.48a) by $[(1 - z)(1 + z)]^{-\mu/2}$ we can shift the indices at $z = \pm 1$ by $-\mu/2$, or shift the index at $z = \infty$ by μ, and obtain the second equation (14.48b). Furthermore, $P\{\cdots\}$ satisfies

$$(1 - z^2)\frac{d^2 w(z)}{dz^2} - 2(\mu + 1)\frac{dw(z)}{dz} + (\nu - \mu)(\nu + \mu + 1)w(z) = 0. \tag{14.49}$$

If z is real and $-1 < z \leq 1$, then when $\nu = n$, $\mu = m$ (n, m are both integers), as we have already learned, the first solution is associated Legendre polynomial (14.23) or (14.24). Exactly as when we derived these equations, we can obtain the equation below by repeating the transformation which leads to Gauss differential equation. However, if z is real and $|z| > 1$, or when it is complex, it is generally defined as

$$P_\nu^\mu(z) = \frac{\Gamma(\nu + \mu + 1)}{2^\mu \mu! \Gamma(\nu - \mu + 1)}$$

$$\times (z^2 - 1)^{\mu/2} F\left(\mu - \nu, \nu + \mu + 1; \mu + 1; \frac{1 - z}{2}\right). \qquad (14.50)$$

We must be careful about that as the same symbol is used.

If $\mu = m$ (a positive integer), we can write the equation taking $\mu \to m$ in Eq. (14.50), as follows:

$$P_\nu^m(z) = (z^2 - 1)^{m/2} \frac{d^m}{dz^m} P_\nu(z). \qquad (14.51)$$

Proof. If Eq. (14.31) is differentiated m times for z, then we get

$$\frac{d^m}{dz^m} P_\nu(z) = \sum_{k=m}^{\infty} \frac{(-1)^m [\nu + 1]_k [-\nu]_k}{k!(k - m)!} \cdot \frac{(1 - z)^{k-m}}{2^k}$$

$$= \frac{(-1)^m}{2^m} \sum_{k=0}^{\infty} \frac{[\nu + 1]_m [\nu + m + 1]_k \cdot [-\nu]_m [-\nu + m]_k}{k! \cdot [m + 1]_k m!} \frac{(1 - z)^k}{2^k}$$

$$= \frac{(-1)^m [\nu + 1]_m [-\nu]_m}{2^m m!} F\left(\nu + m + 1, -\nu + m; m + 1; \frac{1 - z}{2}\right)$$

$$= \frac{\Gamma(\nu + m + 1)}{2^m m! \Gamma(\nu - m + 1)} F\left(\nu + m + 1, -\nu + m; m + 1; \frac{1 - z}{2}\right).$$

Equation (14.50) can be obtained from this, so Eq. (14.51) is true.   □

Similarly, for the independent second solution, we obtain

$$Q_\nu^m(z) = (z^2 - 1)^{m/2} \frac{d^m}{dz^m} Q_\nu(z). \qquad (14.52)$$

Differentiating Eq. (14.33) m times, in the same way as for $P_\nu^m(z)$, produces

$$\frac{d^m}{dz^m} Q_\nu(z) = B\left(\frac{1}{2}, \nu + 1\right) \frac{1}{2^{\nu+1}} \frac{d^m}{dz^m} \frac{1}{z^{\nu+1}} F\left(\frac{\nu + 1}{2}, \frac{\nu}{2} + 1; \nu + \frac{3}{2}; \frac{1}{z^2}\right)$$

$$= \frac{(-1)^m \sqrt{\pi}}{2^{\nu+1}} \frac{\Gamma(\nu + m + 1)}{\Gamma(\nu + \frac{3}{2})} z^{-\nu-m-1}$$

$$\times F\left(\frac{\nu + m + 2}{2}, \frac{\nu + m + 1}{2}; \nu + \frac{3}{2}; \frac{1}{z^2}\right).$$

From this, we obtain

$$Q_\nu^m(z) = \frac{e^{m\pi i} \sqrt{\pi}}{2^{\nu+1}} \cdot \frac{\Gamma(\nu + m + 1)}{\Gamma(\nu + \frac{3}{2})} \cdot z^{-\nu-m-1} (z^2 - 1)^{m/2}$$

$$\times F\left(\frac{\nu + m + 2}{2}, \frac{\nu + m + 1}{2}; \nu + \frac{3}{2}; \frac{1}{z^2}\right). \qquad (14.53)$$

The replacement $m \to \mu$ in this equation is the general expression for general μ.

14.4.2 Recurrence formulae for associated Legendre functions

The following recurrence formula is true for associated Legendre functions $P_\nu^m(x)$, $Q_\nu^m(x)$. Here, we write for functions $P_\nu^m(x)$, but the same equations are satisfied for $Q_\nu^m(x)$ because we do not use, for the proof, specific form of $P_\nu^m(x)$. To keep the explanation simple, let the variable x be real and $-1 < x \le 1$.

- Recurrence formulae for ν

$$\left[(1 - x^2)\frac{d}{dx} + \nu x\right]P_\nu^m(x) = (\nu + m)P_{\nu-1}^m(x) \qquad (14.54a)$$

$$\left[(1 - x^2)\frac{d}{dx} - (\nu + 1)x\right]P_\nu^m(x) = -(\nu - m + 1)P_{\nu+1}^m(x) \qquad (14.54b)$$

are true. We will prove them later, but, by eliminating the differential terms from these two equations, we obtain the following three-term recurrence formula.

$$(2\nu + 1)zP_\nu^m(x) = (\nu + m)P_{\nu-1}^m(x) + (\nu - m + 1)P_{\nu+1}^m(x) \qquad (14.54c)$$

- Recurrence formulae for m

$$\left(\sqrt{1 - x^2}\frac{d}{dx} + \frac{mx}{\sqrt{1 - x^2}}\right)P_\nu^m(x) = P_\nu^{m+1}(x) \qquad (14.55a)$$

$$\left(\sqrt{1 - x^2}\frac{d}{dx} - \frac{mx}{\sqrt{1 - x^2}}\right)P_\nu^m(x) = -(\nu + m)(\nu - m + 1)P_\nu^{m-1}(x)$$
$$(14.55b)$$

are true. We will also prove them later, but eliminating the differential terms from these two equations, we obtain the following three-term recurrence formula.

$$P_\nu^{m+1}(x) - \frac{2mx}{\sqrt{1 - x^2}}P_\nu^m(x) + (\nu + m)(\nu - m + 1)P_\nu^{m-1}(x) = 0. \qquad (14.55c)$$

Proof of Eqs. (14.54a) and (14.54b). Differentiating m times Eq. (14.37a) with replacement of $\nu \to \nu - 1$, we obtain

$$\frac{d^{m+1}P_\nu}{dx^{m+1}} - x\frac{d^{m+1}P_{\nu-1}}{dx^{m+1}} - m\frac{d^m P_{\nu-1}}{dx^m} = \nu\frac{d^m P_{\nu-1}}{dx^m}.$$

Also, by differentiating Eq. (14.37b) m times, we obtain

$$x\frac{d^{m+1}P_\nu}{dx^{m+1}} - \frac{d^{m+1}P_{\nu-1}}{dx^{m+1}} + m\frac{d^m P_\nu}{dx^m} = \nu\frac{d^m P_\nu}{dx^m}.$$

Eliminating $\frac{\mathrm{d}^{m+1}P_{\nu-1}}{\mathrm{d}x^{m+1}}$ from these two equations produces

$$(1-x^2)\frac{\mathrm{d}^{m+1}P_\nu}{\mathrm{d}x^{m+1}} - mx\frac{\mathrm{d}^m P_\nu}{\mathrm{d}x^m} = (\nu+m)\frac{\mathrm{d}^m P_{\nu-1}}{\mathrm{d}x^m} - \nu x\frac{\mathrm{d}^m P_\nu}{\mathrm{d}x^m}.$$

We can obtain Eq. (14.54a) by multiplying by $(1-x^2)^{\frac{m}{2}}$ and reorganizing. Using the fact that this result and P_ν^m satisfy associated Legendre differential equation, we obtain Eq. (14.54b). □

Proof of Eqs. (14.55a) and (14.55b). Using Eq. (14.51),

$$\left(\sqrt{1-x^2}\frac{\mathrm{d}}{\mathrm{d}x} + \frac{mx}{\sqrt{1-x^2}}\right)P_\nu^m(x)$$

$$= \left(\sqrt{1-x^2}\frac{\mathrm{d}}{\mathrm{d}x} + \frac{mx}{\sqrt{1-x^2}}\right)(1-x^2)^{m/2}\frac{\mathrm{d}^m P_\nu}{\mathrm{d}x^m}$$

$$= (1-x^2)^{(m+1)/2}\frac{\mathrm{d}^{m+1}P_\nu}{\mathrm{d}x^{m+1}} - mx(1-x^2)^{(m-1)/2}\frac{\mathrm{d}^m P_\nu}{\mathrm{d}x^m}$$

$$\quad + mx(1-x^2)^{(m-1)/2}\frac{\mathrm{d}^m P_\nu}{\mathrm{d}x^m}$$

$$= (1-x^2)^{(m+1)/2}\frac{\mathrm{d}^{m+1}P_\nu}{\mathrm{d}x^{m+1}} = P_\nu^{m+1}.$$

This is Eq. (14.55a).

Rewriting associated Legendre differential equation as

$$\left[\sqrt{1-x^2}\frac{\mathrm{d}}{\mathrm{d}x} - \frac{(m+1)x}{\sqrt{1-x^2}}\right]\left[\sqrt{1-x^2}\frac{\mathrm{d}}{\mathrm{d}x} + \frac{mx}{\sqrt{1-x^2}}\right]P_\nu^m = 0$$

and using Eq. (14.55a) with it, we obtain Eq. (14.55b). □

14.4.3 *Associated Legendre polynomial*

Until now, we have considered $P_\nu^\mu(z)$ and $Q_\nu^\mu(z)$ as generally as possible. However, in many problems in physics and engineering, boundary conditions, etc. mean that they take the forms $\nu = n$, $\mu = m$ (each is 0 or an integer). Let's discuss this in a little more detail from this kind of perspective.

14.4.3.1 *Definition of associated Legendre polynomial*

When $|x| < 1$, associated Legendre polynomial is

$$P_n^m(x) = \frac{\Gamma(n+m+1)}{2^m m!\Gamma(n-m+1)}(1-x^2)^{m/2}F\left(m-n, n+m+1; m+1; \frac{1-x}{2}\right)$$

according to the definition in Eq. (14.23). This is integer-order associated Legendre's differential equation, which is a differential equation to satisfy that. We have already described

$$\frac{\mathrm{d}}{\mathrm{d}x}\left[(1-x^2)\frac{\mathrm{d}P}{\mathrm{d}x}\right] + \left[n(n+1) - \frac{m^2}{1-x^2}\right]P = 0$$

in Eq. (14.17).

14.4.3.2 Normalization orthogonal relation of associated Legendre polynomial

Jacobi's polynomial $P_n^{(\alpha,\beta)}(x)$ defined in Eq. (13.5) is given by Rodrigues' formula, so as shown in Sec. 13.1.2.2, orthogonality is satisfied when $k \neq l$. Associated Legendre polynomial is for $\alpha = \beta = m$ in this case, so it satisfies the orthogonal relation in the same way, and

$$\int_{-1}^{1}(1-x^2)^m \frac{\mathrm{d}^m P_k(x)}{\mathrm{d}x^m}\cdot\frac{\mathrm{d}^m P_l(x)}{\mathrm{d}x^m}\mathrm{d}x = 0$$

is true. Therefore, when $k \neq l$,

$$\int_{-1}^{1} P_k^m(x)P_l^m(x)\mathrm{d}x = \int_{-1}^{1}(1-x^2)^m \frac{\mathrm{d}^m P_k(x)}{\mathrm{d}x^m}\cdot\frac{\mathrm{d}^m P_l(x)}{\mathrm{d}x^m}\mathrm{d}x = 0. \quad (14.56)$$

When $k = l$, we can obtain the normalization integral by using Rodrigues' formula, or through calculation by repeated partial integration.

$$\int_{-1}^{1}[P_k^m(x)]^2\mathrm{d}x = \int_{-1}^{1}(1-x^2)^m\left[\frac{\mathrm{d}^m}{\mathrm{d}x^m}P_k(x)\right]^2\mathrm{d}x$$

$$= \frac{(k+m)!}{(k-m)!}\cdot\frac{2}{2k+1}. \quad (14.57)$$

Chapter 15

Functions Written With Confluent Hypergeometric Functions

15.1 Weber–Hermite differential equations and Hermite functions

15.1.1 *Problems in physics: Harmonic oscillator*

Various functions appear in problems of physics, particularly those involving dynamics and quantum mechanics. An oscillation motion within a potential proportional to the square of the distance from the center is called harmonic motion. The motion of a particle attached to a spring is an example. This is a situation in classical mechanics, but the same kind of system can be considered in quantum mechanics. This is called a harmonic oscillator, and the equation which describes this motion can be given by the following differential equation (Schödinger equation), in one-dimensional space.

$$\left(-\frac{\hbar^2}{2m}\frac{d^2}{dx^2} + \frac{1}{2}m\omega_0{}^2 x^2 \right)\psi(x) = E\psi(x). \tag{15.1}$$

Here, $\hbar$ is the Planck constant h divided by 2π ($\hbar = h/(2\pi)$), m is the mass of the particle, ω_0 is a constant which we will explain later, $\psi(x)$ is a function of the coordinate x, called a wave function, and describes the state of the particle, and E is the energy of that state. We will consider the method of how to solve Eq. (15.1) without further consideration of its physics.

15.1.2 *Weber–Hermite differential equations and series solutions*

Performing variable transformation

$$\zeta = \sqrt{\frac{m\omega_0}{\hbar}}x = \alpha x \quad \lambda = \frac{2E}{\hbar\omega_0}, \tag{15.2}$$

249

Eq. (15.1) is rewritten as

$$\frac{d^2 w(\zeta)}{d\zeta^2} + (\lambda - \zeta^2)\, w(\zeta) = 0, \qquad (15.3)$$

which is called Weber's differential equation.

If we perform a further variable transformation

$$w(\zeta) = e^{-\zeta^2/2} v(\zeta) \qquad (15.4)$$

we can obtain

$$\frac{d^2 v(\zeta)}{d\zeta^2} - 2\zeta \frac{dv(\zeta)}{d\zeta} + 2\nu\, v(\zeta) = 0 \qquad (15.5)$$

as a differential equation for $v(z)$, where ν is

$$\nu = \frac{\lambda - 1}{2}.$$

Equation (15.5) is called Hermite's differential equation. In terms of a harmonic oscillator, this is a meaningful solution if ν is 0 or a positive integer.

In Hermite's differential equation (15.5), choosing a new variable,

$$z = \zeta^2$$

we obtain

$$z\frac{d^2 v(z)}{dz^2} + \left(\frac{1}{2} - z\right)\frac{dv(z)}{dz} + \frac{\nu}{2}v(z) = 0. \qquad (15.6)$$

This differential equation can, of course, be solved with the method of series solution, but here we will reduce it to a confluent hypergeometric differential equation, as we have already discussed. This equation is Kummer differential equation (12.103) with fixing

$$\gamma = \frac{1}{2}, \qquad \alpha = -\frac{\nu}{2}.$$

Thus, the solution of the differential equation (15.6) can be written as a confluent hypergeometric function, and the first and second solutions are

$$v_1(z) = F\left(-\frac{\nu}{2}; \frac{1}{2}; z\right) \qquad (15.7a)$$

$$v_2(z) = z^{1/2} F\left(\frac{1-\nu}{2}; \frac{3}{2}; z\right), \qquad (15.7b)$$

respectively. Getting back to the start, the solutions for Weber differential equation (15.3) are

$$w_1(\zeta) = e^{-\zeta^2/2} F\left(-\frac{\nu}{2}; \frac{1}{2}; \zeta^2\right) \qquad (15.8a)$$

$$w_2(\zeta) = e^{-\zeta^2/2} \zeta F\left(\frac{1-\nu}{2}; \frac{3}{2}; \zeta^2\right), \qquad (15.8b)$$

respectively.

According to Eq. (12.104),[1]

$$F\left(-\frac{\nu}{2};\frac{1}{2};\zeta^2\right) = \sum_{k=0}^{\infty} \frac{\left(-\frac{\nu}{2}\right)\left(-\frac{\nu}{2}+1\right)\cdots\left(-\frac{\nu}{2}+k-1\right)}{k!\frac{1}{2}\left(\frac{1}{2}+1\right)\cdots\left(\frac{1}{2}+k-1\right)}\zeta^{2k}$$

$$F\left(\frac{1-\nu}{2};\frac{3}{2};\zeta^2\right) = \sum_{k=0}^{\infty} \frac{\left(\frac{1-\nu}{2}\right)\left(\frac{1-\nu}{2}+1\right)\cdots\left(\frac{1-\nu}{2}+k-1\right)}{k!\frac{3}{2}\left(\frac{3}{2}+1\right)\cdots\left(\frac{3}{2}+k-1\right)}\zeta^{2k}.$$

This means that the behavior of $w_1(\zeta)$ and $w_2(\zeta)$ at $\zeta \to \infty$, for general ν

$$w_1(\zeta) \sim e^{-\zeta^2/2}e^{\zeta^2}, \qquad w_2(\zeta) \sim e^{-\zeta^2/2}\zeta e^{\zeta^2}.$$

Both of them do not satisfy the physical requirements $w_1(\zeta) \to 0$ and $w_2(\zeta) \to 0$ ($\zeta \to \infty$), which relates to the requirement to bound states in quantum mechanics that the wave function $\psi(x)$ is square integrable. On the other hand, when $\nu = n$ (0 or an integer), this series terminates in a finite term, and can be a physically acceptable solution.

15.1.3 *Hermite polynomials*

In the previous section, we explained that $\nu = n$ (0 or an integer) is necessary for a solution of a harmonic oscillator in Eqs. (15.8a) and (15.8b). In that case,

$$w_1(x) = e^{-x^2/2}F\left(-\frac{n}{2};\frac{1}{2};x^2\right)$$

$$= e^{-x^2/2}\sum_{k=0}^{\infty} \frac{\left(-\frac{n}{2}\right)\left(-\frac{n}{2}+1\right)\cdots\left(-\frac{n}{2}+k-1\right)}{k!\frac{1}{2}\left(\frac{1}{2}+1\right)\cdots\left(\frac{1}{2}+k-1\right)}x^{2k}$$

$$w_2(x) = e^{-x^2/2}xF\left(\frac{1-n}{2};\frac{3}{2};x^2\right)$$

$$= e^{-x^2/2}x\sum_{k=0}^{\infty} \frac{\left(\frac{1-n}{2}\right)\left(\frac{1-n}{2}+1\right)\cdots\left(\frac{1-n}{2}+k-1\right)}{k!\frac{3}{2}\left(\frac{3}{2}+1\right)\cdots\left(\frac{3}{2}+k-1\right)}x^{2k}.$$

Accordingly, if n is 0 or an even integer, $w_1(z)$ becomes a polynomial of finite terms up to $k = n/2$, and if n is an odd integer, $w_2(z)$ becomes a

---

[1]Note that if $\nu = 0$ then $[0]_0 = 1$.

polynomial of finite terms up to $k = (n-1)/2$. Thus, we obtain

$$
\begin{aligned}
\left[w_1(x)\right]_{n=2m} &= e^{-x^2/2} F\left(-m; \frac{1}{2}; x^2\right) \\
&= e^{-x^2/2} \sum_{k=0}^{m} \frac{(-m)(-m+1)\cdots(-m+k-1)}{k!\frac{1}{2}\left(\frac{1}{2}+1\right)\cdots\left(\frac{1}{2}+k-1\right)} x^{2k} \quad (15.9a)
\end{aligned}
$$

$$
\begin{aligned}
\left[w_2(x)\right]_{n=2m+1} &= e^{-x^2/2} x F\left(-m; \frac{3}{2}; x^2\right) \\
&= e^{-x^2/2} \sum_{k=0}^{m} \frac{(-m)(-m+1)\cdots(-m+k-1)}{k!\frac{3}{2}\left(\frac{3}{2}+1\right)\cdots\left(\frac{3}{2}+k-1\right)} x^{2k+1}.
\end{aligned}
$$
$$(15.9b)$$

(One should notice, when $m = 0$, the footnote *1 in the previous page.) Comparing these with the definition of Hermite polynomial (13.20), the even and odd values of n correspond to w_1, w_2 respectively, producing the following result.

$$
H_{2m}(x) = (-1)^m \frac{(2m)!}{m!} F\left(-m; \frac{1}{2}; x^2\right) \tag{15.10a}
$$

$$
H_{2m+1}(x) = 2(-1)^m \frac{(2m+1)!}{m!} x F\left(-m; \frac{3}{2}; x^2\right). \tag{15.10b}
$$

Figure 15.1 shows the behavior of Hermite polynomials.

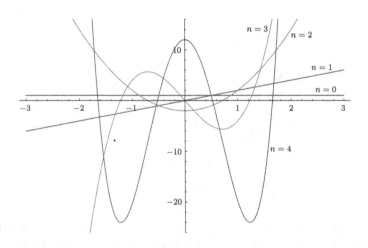

Fig. 15.1 Behavior of Hermite polynomials $H_n(x)$.

15.1.4 Orthogonality and normalization integrals of Hermite polynomials

We have already shown orthogonality in Eq. (13.23). For normalization integrals, consider the case of $f = g = H_n$ in Eq. (13.21) and perform partial integrations repeatedly:

$$\int_a^b e^{-x^2} |H_n(x)|^2 dx = (-1)^n \int_{-\infty}^{\infty} H_n(x) \frac{d^n e^{-x^2}}{dx^n} dx$$

$$= 2^n n! \int_{-\infty}^{\infty} e^{-x^2} dx = 2^n n! \sqrt{\pi}. \qquad (15.11)$$

15.1.5 Generating function of Hermite polynomials

The generating function expansion of Hermite polynomials is given as

$$e^{2xt-t^2} = \sum_{n=0}^{\infty} \frac{H_n(x)}{n!} t^n, \qquad (15.12)$$

and e^{2xt-t^2} is the generating function of Hermite polynomials.

Proof. Applying Goursat's theorem and the definition Eq. (13.19) to Hermite polynomials, we obtain

$$H_n(x) = (-1)^n \frac{e^{x^2} n!}{2\pi i} \oint_C \frac{e^{-z^2}}{(z-x)^{n+1}} dz. \qquad (15.13)$$

The contour path C goes around $z = x$ in the positive direction. The variable transformation $x - z = t$ transforms the contour path C to a contour path C' going around $t = 0$ in the positive direction, and we obtain

$$H_n(x) = \frac{n!}{2\pi i} \oint_{C'} \frac{e^{2xt-t^2}}{t^{n+1}} dt = \left[\frac{d^n}{dt^n} e^{2xt-t^2} \right]_{t=0}. \qquad (15.14)$$

From this, we can obtain Eq. (15.12). $\qquad \square$

Example 15.1. For an Hermite polynomial, it is not difficult to directly show Eq. (15.12) from the definition (13.19). Readers should try it by themselves.

15.2 Laguerre differential equations and Laguerre functions

15.2.1 Problems in physics: Hydrogen atom

The problem of a charged particle (electron) moving in a Coulomb potential centered at $r = 0$, such as an electron in a hydrogen atom, is determined by the Schrödinger equation

$$\left(- \frac{\hbar}{2m} \triangle - \frac{e^2}{4\pi\epsilon_0 r} \right) \psi(r, \theta, \phi) = E\psi(r, \theta, \phi). \qquad (15.15)$$

$\triangle$ is a three-dimensional Laplace operator (Laplacian), $-e$ is the charge on an electron, and ϵ_0 is the dielectric constant of vacuum.

Here, we can assume

$$\psi(r, \theta, \phi) = R_{\gamma lm}(r) Y_{lm}(\theta, \phi). \qquad (15.16)$$

Where, γ is a parameter (a quantum number) that characterizes solutions other than l, m, and must be determined later. $R_{\gamma lm}(r)$ must satisfy the equation

$$\frac{1}{r^2} \frac{d}{dr} \left(r^2 \frac{d}{dr} R_{\gamma lm}(r) \right) + \left[\frac{2m}{\hbar^2} \left(E + \frac{e^2}{4\pi\epsilon_0 r} \right) - \frac{l(l+1)}{r^2} \right] R_{\gamma lm}(r) = 0. \quad (15.17)$$

From this equation, we understand that although we wrote $R_{\gamma lm}(r)$, it does not depend on m of $Y_{lm}(\theta, \phi)$, and henceforth we can write $R_{\gamma l}(r)$.

$$\alpha^2 = \frac{8m|E|}{\hbar^2}, \qquad \lambda = \frac{2me^2}{4\pi\epsilon_0 \alpha\hbar^2} = \frac{e^2}{4\pi\epsilon_0 \hbar} \left(\frac{m}{2|E|} \right)^{1/2} \qquad (15.18)$$

and if we let $\rho = \alpha r$, the above equation becomes

$$\frac{1}{\rho^2} \frac{d}{d\rho} \left(\rho^2 \frac{d}{d\rho} R_{\gamma l}(\rho) \right) + \left[\frac{\lambda}{\rho} - \frac{1}{4} - \frac{l(l+1)}{\rho^2} \right] R_{\gamma l}(\rho) = 0. \qquad (15.19)$$

Furthermore, if we perform a variable transformation to

$$R_{\gamma l}(\rho) = e^{-\rho/2} \rho^l w_{\gamma l}(\rho)$$

the equation that $w(\rho)$ should satisfy is

$$\rho \frac{d^2 w_{nl}(\rho)}{d\rho^2} + [2(l+1) - \rho] \frac{dw_{nl}(\rho)}{d\rho} + (n - l - 1) w_{nl}(\rho) = 0. \qquad (15.20)$$

Here, λ must be an integer such that $\lambda = n \geq l + 1$. At the same time, we replace an unknown parameter γ with n. The fact that λ must be an integer of this kind implies that the behavior of $R_{\lambda l}(\rho)$ when $\rho \to \infty$ requires that R must be a polynomial of finite terms rather than an infinite series as imposed in the case of the harmonic oscillator.

Here we have shown the reason for the appearance of the differential equation (15.20). Further discussion on hydrogen atoms can be found in standard textbooks on quantum mechanics.

15.2.2 Laguerre differential equations or associated differential equations, and series solutions and polynomial solutions

Replacing $2l + 1 \to m$, $n + l \to n$ in Eq. (15.20), we get

$$x\frac{\mathrm{d}^2u(x)}{\mathrm{d}x^2} + (m + 1 - x)\frac{\mathrm{d}u(x)}{\mathrm{d}x} + (n - m)\,u(x) = 0. \tag{15.21}$$

This is called associated Legendre differential equation, and its solution (associated Laguerre polynomial) is written as $L_n^m(z)$. This differential equation can be solved with the method of series solution, and we obtain

$$L_n^m(x) = (-1)^m\frac{n!}{(n - m)!}(e^x x^{-m})\frac{\mathrm{d}^{n-m}}{\mathrm{d}x^{n-m}}(e^{-x}x^n)$$

$$= (-1)^m n! \sum_{k=0}^{n-m}(-1)^k\binom{n}{n - m - k}\frac{x^k}{k!}. \tag{15.22}$$

The associated Legendre differential equation is one of the confluent hypergeometric differential equations when $\alpha = -(n - m)$, $\gamma = m + 1$, so the solution, using the confluent hypergeometric function, can be expressed as

$$L_n^m(x) = (-1)^m\frac{(n!)^2}{m!(n - m)!}F(-n + m; m + 1; x). \tag{15.23}$$

When $m = 0$, in particular,

$$x\frac{\mathrm{d}^2u(x)}{\mathrm{d}x^2} + (1 - x)\frac{\mathrm{d}u(x)}{\mathrm{d}x} + nu(x) = 0 \tag{15.24}$$

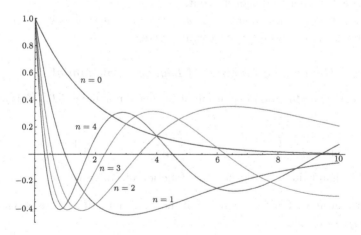

Fig. 15.2 Normalized Laguerre polynomials $\exp(-x/2)L_n(x)$, $n = 0, 1, 2, 3, 4$.

is called Laguerre differential equation, and its solution is also called Laguerre polynomial, which can be expressed as

$$L_n(x) = e^x \frac{d^n}{dx^n}(e^{-x}x^n)$$

$$= n! \sum_{k=0}^{n} (-1)^k \binom{n}{n-k} \frac{x^k}{k!} \tag{15.25a}$$

$$= n! F(-n; 1; x). \tag{15.25b}$$

Figure 15.2 shows their behavior.

15.2.3 Orthogonality and normalization integrals of associated Laguerre polynomials and other polynomials

If $k \neq l$, associated Laguerre polynomials satisfy the orthogonal relation

$$\int_0^\infty e^{-x} x^m L_k^m(x) L_l^m(x) \, dx = 0. \tag{15.26}$$

This relation can easily be shown from Rodrigues' formula, so readers should check the proof.

The normalization integral of associated Laguerre polynomial is

$$\int_0^\infty e^{-x} x^m L_n^m(x)^2 dx = \frac{(n!)^3}{(n-m)!}. \tag{15.27}$$

The proof also uses Rodrigues' formula, and the formula can be partially integrated further to demonstrate it.

For a Laguerre polynomial, the above relation is true as it stands when $m = 0$, so there is no need to explain it here.

15.2.4 Generating function of Laguerre polynomial

The generating function expansion of Laguerre polynomial $L_n(x)$ is as follows.

$$\frac{1}{1-t} \exp\left(-\frac{xt}{1-t}\right) = \sum_{n=0}^{\infty} \frac{L_n(x)}{n!} t^n \qquad (|t| < 1). \tag{15.28}$$

The left-hand side is the generating function of Laguerre polynomials.

Proof. From Goursat's theorem and the definition of Laguerre polynomial (13.14), we obtain

$$L_n(x) = \frac{n! e^x}{2\pi i} \oint_C \frac{z^n e^{-z}}{(z-x)^{n+1}} dz. \tag{15.29}$$

The contour path C goes around $z = x$ in the positive direction on a circle of a radius 1. Here, define

$$z = \frac{x}{1-t}$$

and if we transform the variable from z to t,

$$dz = \frac{x}{(1-t)^2} dt.$$

Also, if it moves (in the positive direction) on a contour path C that goes around $z = x$ in the positive direction, t moves in the positive direction around $t = 0$ on the complex plane. This contour on a complex t plane is written as the contour path C', then

$$L_n(x) = \frac{n!}{2\pi i} \oint_{C'} \frac{e^{-xt/(1-t)}}{(1-t)t^{n+1}} dt = \left[\frac{d^n}{dt^n} \frac{e^{-xt/(1-t)}}{(1-t)} \right]_{t=0}. \tag{15.30}$$

From this, we obtain the generating function expansion (15.28). $\qquad \square$

Example 15.2. For the generating function of Laguerre polynomials, in the same way as for Legendre polynomial, we can obtain this right-hand side by direct calculation from

$$\sum_{n=0}^{\infty} \frac{L_n(x)}{n!} t^n = \sum_{n=0}^{\infty} \frac{t^n e^x}{2\pi i} \oint \frac{z^n e^{-z}}{(z-x)^{n+1}} dz. \tag{15.31}$$

Readers should check this by themselves.

Example 15.3. For associated Laguerre polynomials, we can obtain the integral representation and the generating function expansion in the same way. We will only present the result here, and readers should demonstrate this.

$$L_n^m(x) = (-1)^m \frac{n! e^x x^{-m}}{2\pi i} \oint_{C'} \frac{e^{-t} t^n}{(t-x)^{n-m+1}} dt \tag{15.32a}$$

$$(-1)^m \frac{e^{-xt/(1-t)}}{(1-t)^{m+1}} = \sum_{n=0}^{\infty} \frac{L_{n+m}^m(x)}{(n+m)!} t^n. \tag{15.32b}$$

15.3 Bessel's differential equations and Bessel functions

15.3.1 *Problems in physics: Oscillation of a circular drum*

Example 15.4. Express the oscillation equation of a circular drum $0 \le r \le a$ ($r = \sqrt{x^2 + y^2}$), which is fixed around the periphery, in two-dimensional polar coordinates as

$$x = r \cos \theta, \qquad y = r \sin \theta, \tag{15.33}$$

then assume a separable form of the solution u for r, θ, t as

$$u(r, \theta, t) = R(r)\,\Theta(\theta)\,T(t). \tag{15.34}$$

In the problem of a circular drum, the boundary is $r = a$, so the fixed boundary condition is expressed by

$$R(a) = 0. \tag{15.35}$$

The two-dimensional oscillation equation, expressed in polar coordinates, is

$$\frac{\partial^2 u}{\partial t^2} = c^2 \left(\frac{\partial^2}{\partial r^2} + \frac{1}{r}\frac{\partial}{\partial r} + \frac{1}{r^2}\frac{\partial^2}{\partial \theta^2} \right) u. \tag{15.36}$$

Substituting Eq. (15.34) into Eq. (15.36), we get the following.

$$\frac{1}{T}\frac{d^2 T}{dt^2} = c^2 \frac{1}{R}\left(\frac{d^2}{dr^2} + \frac{1}{r}\frac{d}{dr} \right) R + c^2 \frac{1}{r^2}\frac{1}{\Theta}\frac{d^2 \Theta}{d\theta^2}. \tag{15.37}$$

Each term is a function of only time t, the radial coordinate r, and angle θ, so each must be a constant. From this, the differential equations which should satisfy $T(t)$ and angle $\Theta(\theta)$ are

$$\frac{d^2 T}{dt^2} + \lambda c^2 T = 0 \tag{15.38a}$$

$$\frac{d^2 \Theta}{d\theta^2} + \mu^2 \Theta = 0 \tag{15.38b}$$

respectively. This is a simple differential equation that is easy to solve. On the other hand, substituting these into Eq. (15.37) for $R(r)$ demonstrates that the differential equation

$$\left(\frac{d^2}{dr^2} + \frac{1}{r}\frac{d}{dr} \right) R + \left(\lambda - \frac{\mu^2}{r^2} \right) R = 0 \tag{15.39}$$

must be satisfied.

If we do not consider whether λ is positive or negative and let

$$\rho = \sqrt{\lambda}\, r \tag{15.40}$$

then Eq. (15.39) becomes

$$\left(\frac{d^2}{d\rho^2} + \frac{1}{\rho}\frac{d}{d\rho} \right) R + \left(1 - \frac{\mu^2}{\rho^2} \right) R = 0. \tag{15.41}$$

By the fixed boundary condition for $R(r)$, the domain of θ given by Eq. (15.35) is

$$0 \leq \theta \leq 2\pi, \tag{15.42}$$

but in this case, $\theta = 0$ and $\theta = 2\pi$ must be continuously connected. Therefore, the boundary condition for $\Theta(\theta)$ becomes

$$\Theta(0) = \Theta(2\pi). \tag{15.43}$$

This kind of boundary condition is called a periodic boundary condition.

15.3.2 Bessel's differential equations and expressing solutions with confluent hypergeometric functions

The differential equation in Eq. (15.41)

$$\frac{d^2 w(z)}{dz^2} + \frac{1}{z}\frac{dw(z)}{dz} + \left(1 - \frac{\nu^2}{z^2}\right) w(z) = 0 \tag{15.44}$$

is called Bessel's differential equation. In Example 12.3, we discussed a specific form of Bessel differential equation. It has $z = 0$ as a regular singular point, and $z = \infty$ as an irregular singular point, and has one solution when $\nu = n$ (0 or a positive integer).

Here, if we transform

$$\zeta = 2iz, \tag{15.45a}$$

$$w(z) = e^{-iz} z^\nu u(2iz) \tag{15.45b}$$

then Eq. (15.44) becomes

$$\frac{d^2 u(\zeta)}{d\zeta^2} + \left(\frac{2\nu+1}{\zeta} - 1\right)\frac{du(\zeta)}{d\zeta} - \frac{\nu+\frac{1}{2}}{\zeta} u(\zeta) = 0. \tag{15.46}$$

This is confluent hypergeometric differential equation for $\alpha = \nu + 1/2$ and $\gamma = 2\nu + 1$, so if ν is not an integer, it can be written with the following confluent hypergeometric function.

$$w_1(x) = e^{-iz} z^\nu F\left(\nu + \frac{1}{2}; 2\nu + 1; 2iz\right) \tag{15.47a}$$

$$w_2(x) = e^{-iz} z^{-\nu} F\left(-\nu + \frac{1}{2}; -2\nu + 1; 2iz\right). \tag{15.47b}$$

In particular,

$$J_\nu(x) = \frac{e^{-iz} z^\nu}{2^\nu \Gamma(\nu+1)} F\left(\nu + \frac{1}{2}; 2\nu + 1; 2iz\right) \tag{15.48}$$

is called Bessel's function of order ν. Bessel differential equation is constant when the substitution $\nu \to -\nu$ is performed. This means that $J_{-\nu}(z)$ is another solution. Normally, if ν is 0 or non-integer,

$$N_\nu(z) = \frac{\cos(\nu\pi)J_\nu(z) - J_{-\nu}(z)}{\sin(\nu z)} \tag{15.49}$$

is employed as a solution that is independent of $J_\nu(z)$. This $N_\nu(z)$ is called Neumann function of order ν.

15.3.3 *Bessel functions of integer orders*

If ν is 0 or an integer,

$$J_{-n}(z) = (-1)^n J_n(z) \tag{15.50}$$

and J_n and J_{-n} are no longer independent.

Proof. From Eq. (15.40), we show that

$$
\begin{aligned}
J_{-n}(z) &= \left(\frac{z}{2}\right)^{-n} \sum_{m=n}^{\infty} \frac{(-1)^m}{\Gamma(-n+m+1)} \left(\frac{z}{2}\right)^{2m} \\
&= \left(\frac{z}{2}\right)^{-n} \sum_{m=0}^{\infty} \frac{(-1)^{n+m}}{\Gamma(m+1)(n+m)!} \left(\frac{z}{2}\right)^{2(n+m)} \\
&= (-1)^n \left(\frac{z}{2}\right)^{n} \sum_{m=0}^{\infty} \frac{(-1)^m}{\Gamma(n+m+1)} \left(\frac{z}{2}\right)^{2m} \\
&= (-1)^n J_n(z).
\end{aligned}
$$

$\square$

In that case, if we use Eq. (15.49) and define Neumann function of integer order as

$$N_n(z) = \lim_{\nu \to n} N_\nu(z). \tag{15.51}$$

This is another independent solution. This equation becomes indeterminate of the type $0/0$, so we take the limiting value of the ratio of denominator and numerator, and we get

$$N_n(z) = \frac{1}{\pi} \left[\left(\frac{\partial J_\nu(z)}{\partial \nu} \right)_{\nu=n} - (-1)^n \left(\frac{\partial J_{-\nu}(z)}{\partial \nu} \right)_{\nu=n} \right]. \tag{15.52}$$

This is the second solution, which has already appeared in Eq. (12.43b). The behavior of Bessel functions and Neumann functions of order integer numbers are shown in Fig. 15.3. The reader can demonstrate this as a practice problem.

15.3.4 *Half-integer Bessel functions and spherical Bessel functions*

Bessel functions and Neumann functions of half-odd integer order ν can be expressed, as follows, with elementary functions. These functions appear

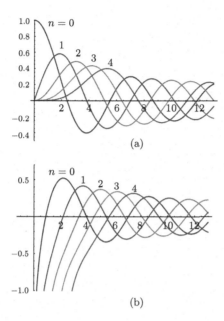

Fig. 15.3　(a) Bessel functions $J_n(x)$ of integer order, and (b) Neumann functions $N_n(x)$.

widely in the application. Specifically, we will write Bessel functions and Neumann functions of half-odd integer orders as follows:

$$J_{n+1/2}(z) = \sqrt{\frac{2}{\pi}} z^{n+1/2} \left(-\frac{d}{z\,dz} \right)^n \frac{\sin z}{z} \tag{15.53a}$$

$$J_{-n-1/2}(z) = \sqrt{\frac{2}{\pi}} z^{n+1/2} \left(\frac{d}{z\,dz} \right)^n \frac{\cos z}{z} \tag{15.53b}$$

$$N_{n+1/2}(z) = (-1)^{n+1} J_{-n-1/2}(z). \tag{15.53c}$$

The solutions to the differential equations

$$\frac{d^2}{dz^2} w(z) + \frac{2}{z} \frac{d}{dz} w(z) + \left[1 - \frac{n(n+1)}{z^2} \right] w(z) = 0 \tag{15.54}$$

can be written down as

$$j_n(z) = \sqrt{\frac{\pi}{2z}} J_{n+1/2}(z) \tag{15.55a}$$

$$n_n(z) = \sqrt{\frac{\pi}{2z}} N_{n+1/2}(z) \tag{15.55b}$$

if one uses variable transformation. The $j_n(z)$, $n_n(z)$ which appear here are called as spherical Bessel function and spherical Neumann function, respectively.

Bibliography

Analysis and complex function theory (general)

Teiji Takagi: *An Introduction to Analysis* (in Japanese), revised edition, (Iwanami Shoten, 1961), first published in 1938.
A classic masterpiece of a textbook, covering the whole of analysis. It includes a well-balanced explanation of complex analysis.

Kanichi Terazawa: *An Introduction to Mathematics for Natural Scientists* (in Japanese), (Iwanami Shoten, 1954).
A textbook on analysis for those studying science and engineering, and for those working in engineering. It covers a wide range of fields.

V. I. Smirnov: *A Course of Higher Mathematics*, Vol. 3 Part 2, (Pergamon Press, 1964).

E. T. Whittaker and G. N. Watson: *A Course of Modern Analysis*, Cambridge Mathematical Library, 4th Edition, (Cambridge University Press, 1952), first published in 1902.

Lars V. Ahlfors: *Complex Analysis*, (McGraw-Hill, 1979).
This book is carefully written, and is easy to understand, despite being written by a mathematician. It is the established classic work, used around the world.

Elliptic functions

A. Hurwitz and R. Courant: *Vorlesungen über allgemeine Funktionentheorie und elliptische Funktionen*, Zweiter Abscnitt, Elliptische Functionen, (Verlag von Julius Springer, Berlin, 1922).

M. Toda: *An Introduction to Elliptic Functions* (in Japanese), Nippyo Library of Mathematics, (Nippon Hyoron Sha Co. Ltd., 2001).
This textbook contains many important examples. The explanation is very easy to understand and the author recommends strongly reading it.

M. Abramowitz, and I. A. Stegun, eds.: *Handbook of Mathematical Functions with Formulas, Graphs, and Mathematical Tables*, Dover Books on Advanced Mathematics, (Dover Publications, 1983).

Ordinary differential equations, hypergeometric functions, and others of complex variables

E. E. Kummer; *Journal für die reine und angewandte Mathematik* **15**, 39–83 (1836).

T. Inui: *Special Functions* (in Japanese), (Iwanami Shoten, 1962).

H. Hochstadt: *The Functions of Mathematical Physics*, Dover Books on Physics, (Dover Publications, 2012).

M. Abramowitz, and I. A. Stegun, eds.: *Handbook of Mathematical Functions with Formulas, Graphs, and Mathematical Tables*, Dover Books on Advanced Mathematics, (Dover Publications, 1983).

Index

Printed in the United States
by Baker & Taylor Publisher Services